How to DNA Test Our Family Relationships?

By

Terrence Carmichael, M.S.
and
Alexander Kuklin, Ph.D.

Contributing authors:
Ed Grotjan, Ph.D.
Robert Allan, Ph.D.
James P. Reape, Attorney at Law

Text Editor:
Emily Balestrini

Ace N Press

HOW TO DNA TEST OUR FAMILY RELATIONSHIPS?

by Terrence Carmichael and Alexander Kuklin

Published by:

AceN Press

Mountain View, CA, USA

First printing 2000

Printed in the United States of America

Library of Congress Cataloging-in-Publication Data

Carmichael, Terry, 1967-

How to DNA test our family relationships / by Terry Carmichael and Alexander Kuklin ; contributing authors Ed Grotjan ... [et al.].-- 1st ed.

p. cm.

Includes bibliographical references and index.

ISBN 0-9664027-1-5 (pbk. : alk. paper)

1. Paternity testing. 2. Medical genetics. 3. DNA--Analysis. I. Kuklin, Alexander Ivanov, 1962- . II. Grotjan, Ed. III. Title.

RA1138 .C37 2000

616'.042--dc21

00-010875

CIP

Direct all inquiries to AceN Press, acenpress@hotmail.com

TABLE OF CONTENTS

CHAPTER **PAGE**

CHAPTER	PAGE
PREFACE	vii
INTRODUCTION	11
PATERNITY TESTING	15
WHAT TO EXPECT IN A PATERNITY TEST?	21
THE HISTORY THAT LED TO UNDERSTANDING HEREDITY	31
UNDERSTANDING THE HUMAN GENOME	39
PCR: AMPLIFYING YOUR GENES	53
DNA FRAGMENT SEPARATION TECHNIQUES USED IN PATERNITY TESTING	61
DNA PROFILING: A UNIQUE METHOD FOR IDENTIFICATION	67
HOW TO READ A PATERNITY TEST REPORT	79
SIBLINGSHIP AND GRAND-PARENTAGE STUDIES	93

SEROLOGICAL METHODS USED TO DETERMINE BIOLOGICAL RELATIONSHIPS 113

PARENTAGE TESTING IN THE UNITED STATES: THE ROLE OF AABB 119

SELECTED RESOURCES 123

GLOSSARY 125

APPENDIX I: A LEGAL PERSPECTIVE 135

APPENDIX II: DNA TESTING RESOURCES FOR YOUR FAMILY 149

INDEX 153

PREFACE

The progress of the Human Genome Project caused an explosion of renewed interest in using DNA technology to study people's ancestry. Scientific discoveries pointing to Africa as the recent birthplace of modern humans used DNA, passed exclusively in a maternal line by the mitochondria, the energy producing organelles in the cell. Researchers traced this DNA to a mitochondrial Eve. To test the theory, scientists studied the Y chromosome, which is passed from father to son. This provided an independent test of the mitochondrial Eve hypothesis. Scientists found the genetic trail leading to Adam, and it pointed to the same time and place where mitochondrial Eve lived.

Services based on DNA technology, which are still being refined in research laboratories, are in great demand by consumers who are eager to read the genetic book of their ancestors and find their true origins. The avalanche of information and speed of delivery available through the Internet, coupled with guaranteed anonymity and privacy offered by e-commerce genetic companies, allows those interested to easily order such tests on-line.

This book is a short reference guide for the use of DNA technology in testing family relationships. Its objective is to outline the procedures involved in DNA testing. Each topic mentioned here is a scientific discipline by itself. The authors do not cherish the ambition to describe all possible details and discoveries in this field. Our aim is to spark interest in the reader for further studies and discussions.

Each book chapter covers a step of the DNA testing procedure as shown in Figure P-1. The simplest test for studying

Figure P-1

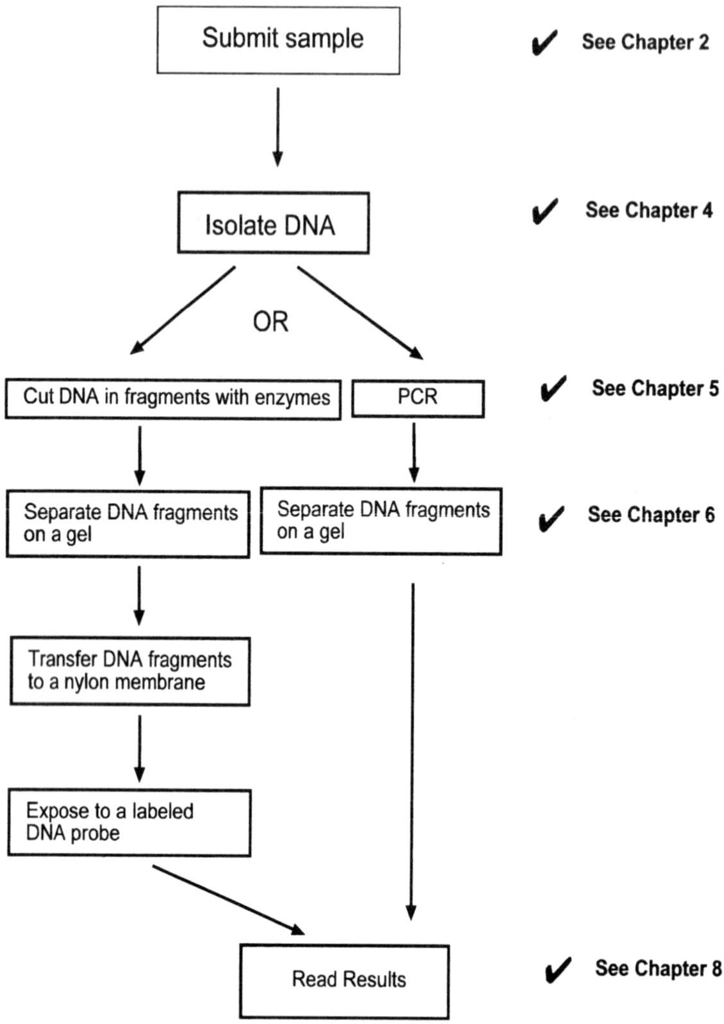
Submit sample
See Chapter 2
Isolate DNA
See Chapter 4
OR
Cut DNA in fragments with enzymes
PCR
See Chapter 5
Separate DNA fragments on a gel
Separate DNA fragments on a gel
See Chapter 6
Transfer DNA fragments to a nylon membrane
Expose to a labeled DNA probe
Read Results
See Chapter 8

family relationships has been the paternity test and therefore, it occupies considerable part of our discussions. This test has been perfected during the last few years, and is efficiently standardized. It starts with extraction of DNA from biological material. The DNA sample is either cut into pieces with enzymes or certain DNA regions are amplified. These DNA regions do not code for proteins. They are represented by different numbers of repeated segments in the region or piece of DNA examined. The obtained DNA fragments are placed in a gel matrix to be separated from each other by size. The bands of interest are resolved and compared for matching as shown in Figure P-2. The child (C) will share one

Figure P-2

Bands on autoradiograph (X-ray film) or a gel picture appear similar to the picture below:

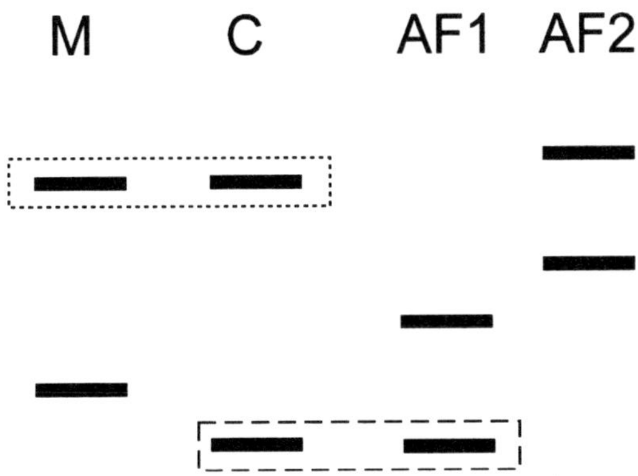

band with the biological mother (M) and one band with the alleged father (AF1), who is the biological father in this example. The alleged father 2 (AF2) shares no bands with the child and therefore, is excluded from being the biological farther. This simple procedure is the basis of DNA testing.

After finishing this book you will know the answers to the following questions:

- What is involved in DNA testing family relationships?
- How are distant family relationships established with DNA technology?
- How can DNA technology help you understand better your ancestry?

We believe that with knowledge about your DNA inheritance you will be able to embrace the notion for your *Global Identity*. We humans are all so different as individuals and at the same time so similar in terms of our ancient DNA ancestry. Now coupled with the ever increasing wealth of knowledge extracted fom DNA research, the differences between individuals can be examined to give insight to one's own ancestrial history.

INTRODUCTION

At 22,000 feet, on the summit of Cerro Llullaillaco, in the Andes, peaks the highest archeological site known to man. In 1999 a group of archaeologists, led by Johan Reinhard, were there fighting with cold and wind waiting for a break from the harsh elements to continue their difficult work excavating Inca ruins. Finally, the fierce wind stopped, and more than five feet down at the excavation site they found the frozen body of an Inca ritual sacrifice. This richly wrapped body belonged to a boy about eight years old. Subsequently, the group uncovered two other mummies of female children a few yards away.

One was of a 14-year-old girl that had been well preserved. "Later, while we were preparing to put the mummy into the freezer at Catholic University in Salta, we examined her more closely," writes Reinhard. "A hush fell over the room as I removed her cloth, revealing her face. For an instant time stopped while I stared in awe at her beauty. Her hair was stylishly braided, and she looked as if she were sleeping. But it was her hands, perfectly life like, that affected me most."

These mummies had survived five centuries to reveal the secrets of Inca human sacrifice rituals. But why were they sacrificed and were they related to each other? To answer these questions, the scientists took samples from their frozen bodies and extracted their mitochondrial DNA. The samples showed that the three children were not siblings, nor were they closely related on the maternal side. But this is just where the use of their DNA begins. The aquired DNA information is now making it possible to identify modern-day relatives and follow their lineage.

Unlike the well preserved Inca children's bodies, the remains of the Russian monarch family Romanov were in the ground for over 70 years after Tsar Nicholas II, his wife, and three daughters were killed by the Bolsheviks in Russia during the revolution of 1917. The family members were brutally shot and buried in a mass grave. Their bones were the only biological material which led scientists to the conclusion that the remains found in the burial place really belonged to the royal family. Tsar Nicholas II was identified by a comparison of the mitochondrial DNA sequence from the bones in the grave with that of the great-great-great granddaughter and great-great grandson of the grandmother of Tsar Nicholas II, Louise of Hesse Cassel.

DNA testing can resolve mysteries of the past, and point to wise decisions, which can help people to cope with the present and plan for the future. However, results from testing personal family relationships with the precision of new molecular biology technique requires psychological sobriety. One should be prepared to deal with such results in a educated and emotionally controlled manner. DNA based tests for establishing family relationships represent a very powerful time machine, allowing us to uncover truths and rewrite history.

Every day thousands of people are using DNA based methods to determine their biological relationships with other members of their families, tribes or clans. There are several reasons why someone would be interested in determining their relationship to another person, such as immigration, resolving legal guardianship, paying child support, property inheritance, awareness of genetic concerns, or just to solve a family dispute regarding someone's relationship.

Questions of paternity or maternity are easy to resolve with molecular biology. A sample of a few cells scraped from the inside of the cheek, hair follicles, or a few drops of blood is all that is needed to solve the riddle of our tru biological relationships.

This book will attempt to address various questions such as:

- ⇨ How can we track our family relationships by DNA testing?
- ⇨ What is inheritance and DNA?
- ⇨ What is the difference between mitochondrial and genomic DNA?
- ⇨ Can we trace back our relatives who have died hundreds of years ago?
- ⇨ Can we establish relationship from body remains that are not well preserved?
- ⇨ What are some of the legalities regarding paternity?
- ⇨ What technologies are used in DNA tests for establishing family relationships?

All of the above questions and more lead to one main inquiry, and that is, "How to DNA test our family relationships." The process of testing family relationships involves learning some technical language and consulting with professionals in the field. This book is a short practical guide for everyone interested in testing their genetic relationships with the use of advanced molecular biology, or DNA based tests. It is easier to understand the reasoning for particular tests when family relationship concerns are demonstrated, such as what was involved in the following situation endured by the Steel brothers.*

Raised side-by-side, Sam and John Steele suspected they might have different fathers. They hadd harbored this unpleasant suspicion from their early teens, and even talked about it when they were in their 30's. Now in their 50's, they have access to technology that can finally determine if indeed they were fathered by the same man, or if perhaps one of them had been switched at birth. Unfortunately, both parents were killed in car accident in 1996; however the brothers can still conduct a full-siblingship

---

*In this and other examples used throughout the book, names and particulars have been changed to protect the parties involved.

analysis to assist in answering their question. Soon after requesting a full-siblingship analysis, Sam and John received a home specimen collection kit to use for collecting their DNA specimens. These kits are designed to collect DNA for analysis by scraping a swab on the inside of the cheek. After collecting the specimens, they sent the samples directly to the lab for analysis.

At the lab, genomic DNA was extracted off of the swabs, and discrete regions of their DNA were amplified by a technique called PCR. Subsequently the results were completed, the probabilities were calculated, and the results demonstrated >99.9% probability that they were full-siblings. This was a pleasant and rewarding surprise to both Sam and John.

One of the brothers shared his feelings after completion of the test, "If it weren't for this test I would have gone through life thinking that one of us was not from this family and that my mother had been dishonest with our father."

Biological relationship testing is a rapidly growing practice that mixes genealogy with DNA based testing, archaeology, emotions, and the truth about an individual's existence. These new applications are changing the way we look at ourselves, our family relationships, our origin and future. Using the powerful methods of DNA fingerprinting, we are capable of accurately and conclusively answering questions needed to solve crimes, determine adultery, construct family trees, resolve immigration issues and identify people. It is our simple, but unique DNA sequences that are truly particular only to one's self and no one else. It is the Double Helix that can tell you what you are, where you come from, where you have been and what you may have done. Furthermore, unlike the standard fingerprint, used in the more traditional method of identifying people, you cannot live without DNA or alter it to disguise yourself. It is found in virtually every cell in your body and can be preserved for a long time after a person has passed away.

Chapter 1

PATERNITY TESTING

PATERNITY TESTING AND SOME STATISTICS

A paternity (or parentage) test consists of laboratory procedures carried out on biological samples obtained from individuals who are involved in a putative family tree. The objective is to determine if an alleged father could be excluded from being the true biological father. Conclusions are drawn based on the principal of biological inheritance, discussed further in Chapter 3. The child should inherit half of the characteristics (DNA markers) in his/her genome from each of his/her true parents.

In most cases, paternal tests are necessary for single mothers who are unable to raise their children without financial assistance from the father. These women need to apply for federal aid. The "Aid to Families with Dependent Children" qualification requires that the mother identifies the biological father. Additionally, many men will seek protection from such claims by requesting an independent paternity test.

The following statistics speak well of the growing awareness and use of paternity tests in the United States. In 1996 a total of 6.24 million pregnancies included 3.89 million live births. The pregnancy rate for unmarried women was 96 per 1,000, which is just 18% lower than the rate for married women (113 per 1,000). The period 1980-95 is characterized by relative stability of live births among married women, which accounts for about three-quarters. In contrast, the proportion of pregnancies among unmarried women ending in live births has increased from 33% in 1980 to 47% in 1995. According to the National Vital Statistics Reports, the increase in non-marital birth rates is due to decline in "shotgun marriage". Additionally, it is contributed to the trend of cohabitation of unmarried women. In the period 1980-84, 29% of births to unmarried women were to cohabiting women.

The total volume of parentage cases has increased during the last few years. In 1998 the number of cases performed at accredited laboratories reached 250,000. This could be attributed to the marketing of companies and increased public awareness, as well as to the improvement of DNA testing and the credibility of DNA technology in court.

DO YOU NEED A PATERNITY TEST OR IS THERE ANOTHER TEST THAT BETTER SUITS YOUR NEEDS?

The most important step for an individual to start a process of DNA studying his or her genealogy or to test a child for paternity/maternity relationships is the decision whether such a test is really needed. This commitment is sometimes psychologically exausting because of the unknowns involved in the outcomes of the test. In many instances DNA testing is an absolute necessity.

The commitment to request a DNA test on your family relationships is challenging because of its unpredicted outcome.

Lisa Barry went to court seeking financial support from a man who supposedly was the father of her son. Her attorney requested the paternity test. Lisa was desperate to get the case resolved because of her financial situation. The alleged father agreed to a test. However, it was a big disappointment to Lisa when she received the results indicating that this man was not the father of her son.

Immigration attorneys know very well how the power of a DNA test can support a case with the objective to establish a paternity/maternity relationship for legal admittance into a country. More interesting, and difficult to justify, are the cases when a claimant seeks admittance to a country due to a political situation in his/her native land. However, in numerous cases the real basis for immigration are poor economical conditions, and the political claim is a way to circumvent the law. In such cases, a genetic test always comes conveniently depending on which side of the court bench one is standing. Legal admission to a country based on a family relation or ethnicity is sometimes granted only after the results from a genetic test prove their claim.

A large number of people are starting to look back into their genealogy and unraveling the past can turn out very challenging sometimes. Late one night a telephone call came through from a man seeking answers for the deep-rooted questions of a lifetime. The man identified himself as Henry J. Stanley, a 45 year old, with five other brothers and sisters. All his siblings were born under the Stanley surname, and grew up believing that Steven J. Stanley was their father. Steven had died 4 years earlier, and samples from his remains were not easily obtainable.

Henry had spent a great deal of time studying his family history. He had traveled to Germany, France, and Scotland tracing his roots associated with the Stanley name. His research was published in a book on the Stanley Family. Aside from all the difficulties he had tracing his roots, someting else was bothering him: He didn't believe that he is his father's biological child. He was looking to use the power of DNA testing to validate if his feelings were correct.

After a long discussion regarding his case, he learned about the availability of a siblingship analysis. Furthermore, by testing three siblings, instead of just two, and also including his mother's samples, the test would be quite revealing, and could even lead to a complete exclusion of one sibling if the other two siblings are certain to have the same biological father. Henry was very excited and eager to see if the others would participate in the study.

A couple of days later Henry called again. He had convinced his two sisters and mother to participate in a full-siblingship analysis. After the test was completed and the results were transcribed, it was time let Henry know that one of his other two siblings did not match the other two! Surprisingly, Henry took the news very well. In fact, he told us that while waiting the long three weeks for the results, he had many opportunities to talk to his mother, Edith Stanley. He said that Edith had shared with him that his sister, Martha, might not be his biological sister, and that she may be the one who was not a Stanley.

Now, Henry's suspicions were still strong, and he wanted to do anything and everything he could to determine if he were a Stanley. He mentioned that the only living Stanley on his father's side was a cousin, Henry N. Stanley, whom he had found while searching for the Stanley roots. A family tree analysis indicated that an Y-chromosome test would suggest if they both were descendents of the same paternal lineage. However, this would not be conclusive evidence either way the results came out, but it could provide suggestive evidence.

Henry's cousin, Norbert, was interested in performing the Y-chromosome analysis. His participation yielded results indicating that they were not from the same paternal lineage. This was not very good news to Henry, and he impetuously decided to exhume Steven Stanley's grave. While researching other alternatives, Henry discovered that a Veterans hospital in Chicago might have had some specimens of Steven Stanley's in storage. This would have facilitated the process, because exhuming a body can be a very difficult process to accomplish due to legal restrictions and procedures, as well as to emotional reasons. Unfortunately, specimens were not available at the

Veteran's hospital, and Henry ended up falling back on the idea to exhume Steven Stanley's grave.

During the discussions Henry was having with his mother regarding his paternal biological relationship, she admitted to Henry that it was actually possible that all of his siblings were not born to Steven Stanley at all, and that in fact Steven Stanley may have been sterile. Henry was very surprised to hear this coming from his mother, but as a true researcher, took the news objectively, and was determined to find his paternal roots. With a little graceful tact, Henry was able to get the name of the man that his mother believed was his true biological father, and also the biological father of most his other siblings. Edith named her long-time employer: Bob McCarthy.

Henry immediately started searching for Bob McCarthy. It didn't take him long to find out that he had died just 5 years earlier. He looked at his will and was able to come up with names that could help him possibly find other descendants of Bob McCarthy who might be willing to undergo DNA testing.

Henry found a man named Bob McCarthy Jr., Bob's son, and possibly Henry's half-brother. Henry persisted in getting Bob Jr. to participate in a Y-chromosome analysis to see if they shared the same paternal lineage. The DNA test examined seven different loci on the Y-chromosome between Henry and Bob Jr., finding that both sets matched identically. Henry was convinced that he had found his paternal parent, Bob McCarthy, and a new half-sibling, Bob Jr. Although Henry did not know his biological father, his biological family tree grew a new branch. And knowing how interested Henry is in uncovering his roots, it is for sure that his digging will not stop here. Last heard, Henry was flying to Indiana to meet the new members of his biological roots.

Checklist for requesting a paternity or distant family relationship genetic test?

YES	NO	
❑	❑	Do I understand completely what is involved in such a test and the possible outcomes?
❑	❑	Will the test benefit me materialistically or emotionally?
❑	❑	Will I continue keeping the same good relations with my family members after such a test?
❑	❑	Will I be able to continue performing my professional and personal duties in the case of a disappointing or unexpected result?

If you answered NO to any of these questions you should **not** persue DNA testing.

Chapter 2

WHAT TO EXPECT IN A PATERNITY TEST?

By H. Edward Grotjan, Ph.D.

What should one expect from a paternity test? 'Either outcome' is the answer. Paternity tests are performed when there is doubt in a person's mind about a true biological relationship to another person. That uncertainty translates to an average exclusion rate of 30 to 33%, meaning that the alleged father is proven not to be the biological father in about

Q: What should I expect from a paternity test?

A: When you commit to performing a parentage test, you should prepare yourself for all possible outcomes.

one-third of the paternity tests performed in the United States each year. When you commit to performing a parentage test, you should prepare yourself for all possible outcomes. The child's mother is the person most likely to know who the child's biological father is, but she may not know for sure.

Mr. Miller brings his 8-year old son for parentage testing, stating that the child's mother (his former wife) had an affair about the time the child was conceived, and that he has always had concerns about whether he is the child's father. Nonetheless, he loves his son and has paid child support during the five years that he and the former wife have been divorced. But the question of parentage has bothered him since the child was born and he decides to test the child to put the question at rest. The test is conducted and Mr. Miller is excluded as the child's father (i.e., he is not the father). Mr. Miller now finds himself in dealing with another set of circumstances. He loves his "son" and must now decide what to do with the parentage information.

Alternatively, Mr. Miller could have been proven to be the biological father relieving years of anxiety. What would you do in such a situation? For good or bad, DNA is not the only factor governing humans' reasoning and conduct. There is much beyond that!

Ms. Usher, a potential grandmother brought in two young adults, both 17 and in high school, along with a new baby, for a paternity test Each party involved in the test, including the baby, had a different last name. The alleged father and mother were not married and the child's last name corresponded to an alleged father who was not being tested. The mother of the child confided that she was sure the alleged father being tested was NOT the biological father of the baby. But the genetic analysis showed otherwise. The alleged father was included (not excluded) among potential fathers of the child. The calculated Paternity Index suggested that it was very likely (>99%) that the alleged father **was** the biological father of the baby. When presented with the results, Ms. Usher was delighted and stated she would be committed to helping her son and his girl friend raise her grandson! Ms. Usher says she had a strong feeling that her son was the father of the child and wanted scientific evidence so that she could help her son plan an appropriate course of action.

The outcome of this test could also have been different.

Ms. Usher's son could have been excluded as the biological father of the child.

The bottom line in a paternity test is that an alleged father is either **excluded** or **not-excluded** as a potential biological father of a child. If the alleged father is not excluded, he is said to be included among the persons who could have fathered the child. This may sound a little confusing. The exclusion part is relatively straightforward. However, the inclusion (not excluded) part may be more difficult to comprehend. Why can't the laboratory simply state that the alleged father is the biological father of the child? From a purely scientific or theoretical perspective, there is always a remote chance that some other person with a similar genetic profile could have fathered the child. Thus, when the alleged father's genetic markers (see further text in the book) match those of the child, paternity testing laboratories usually state that a person is not excluded (which is equivalent to saying that he is included) and will make a statement regarding the degree of certainty. If you remember, the mother in this case was sure that the alleged father not being tested was the true biological father. Did the results from this test exclude him? The answer to this is NO. The other alleged father should also be tested.

This brings up another interesting point called *Presumption of Maternity*. The mother is legally recognized as child's mother unless the genetic analysis proves otherwise. In reality, cases where the mother does not match the child are very rare. Hospitals exert considerable effort to make sure that each newborn is matched with its mother. From a practical perspective, *Presumption of Maternity* is logical and rational. From a legal perspective, it is possible to state who the mother of a child is with certainty—that is, without the "not excluded" qualifier. But it is not legally appropriate to make a similar absolute statement regarding paternity. Alleged fathers are legally **presumed** to be the biological father of a child when the level of statistical certainty exceeds some threshold that is set by law. Presumption of maternity also has consequences in the way that the child's alleles are assigned to have been inherited from its mother and father (see below).

LEGALLY ADMISSIBLE VERSUS NON-ADMISSIBLE TESTS

Paternity tests can be divided into two general categories: legally admissible and non-admissible. Many laboratories offer non-admissible tests at moderate cost. In non-admissible tests, specimens are usually collected at home by the parties involved, and sent to a laboratory for analysis. The results are ordinarily returned as a letter of opinion that carries little legal weight, and although it may be introduced as evidence in a court of law, it is not strongly supported. Some laboratories do not even put the names of the parties tested on the report to avoid any attempt to misuse the results. This type of test is suitable for persons who just want to know for their own information. However, if the outcome is different than anticipated and one of the parties decides to take legal action, the entire paternity test may need to be performed under a different set of circumstances that are more strongly supported.

Legally admissible tests are performed under a set of conditions, including *Chain of Custody*, designed to make sure that the parties tested are positively identified and that specimen collection, handling and analysis are performed by a neutral third party. The term *Chain of Custody* refers to a protocol that documents how the DNA specimens are collected, handled, transported and stored. This written *Chain of Custody* delineates who had responsibility for the samples from the time they were collected until they were destroyed. *Chain of Custody* protocols correspond to how evidence would be handled in a criminal case. The overall concept is to provide evidence that the specimens are from the appropriate persons, and that they have not been tampered with. The persons who collect, witness, transport, handle, analyze and report the results of the test must be independent, third parties with no interest in the outcome of the test. If you are involved in a legally admissible paternity case, you should expect to be positively identified at the time the specimens are collected.

Simply stated, the protocols currently used by paternity testing laboratories today are designed to make sure that substitutions do not occur. Paternity testing laboratories now go to great lengths to identify the parties being tested. Their primary responsibility is to perform the test is an unbiased and accurate manner. Their primary concern is to obtain completely reliable data and reach accurate conclusions because the outcome of each paternity test is extremely important to all parties involved.

Positive identification usually includes having adults present one or more forms of picture identification that may be photocopied (Fig. 2-1). The person who brings the child may be asked to present the child's social security card and/or birth certificate. Each party tested should expect to be photographed and fingerprinted. Each adult should expect to sign a statement affirming that the information given is true and accurate as well as a statement giving the laboratory permission to perform the analysis. Many times, appointments are scheduled so the mother can view the alleged father's photograph. Some companies scan the photographs and include them as integral part of the report to generate a visual record of the parties tested.

When providing information, you can also expect to be asked questions that may affect the outcome of the test. You will be questioned if you have ever had a bone marrow transplant or a recent blood transfusion. If so, there is a remote chance that DNA from the donor may be present in your blood. This does not affect the outcome from buccal swab draws. You will also be asked your race. Discrete racial groups or populations have a different distribution (frequency) of alleles at each genetic locus (a genetic marker). The inquiry regarding race allows the laboratory to use appropriate frequency data when calculating the Paternity Index for an alleged father that is not excluded. Brothers and father-son pairs, in particular, are likely to have similar genetic profiles, and this makes paternity testing more difficult if the alleged fathers are related.

Figure 2-1

Chain-of-custody Form (Optional)
(To be completed by the collector)

To establish a chain-of-custody for the specimens (for a case that may be submitted as evidence in a court of law this is required) the specimen collector should completely fill out everything on this form. Pictures, IDs and fingerprints should be taken from all individuals being tested. Since the law varies from state to state for genetic based tests, please consult an attorney in the state of your jurisdiction on the appropriate method and guidelines to follow when submitting, collecting, and shipping samples that are to be used as evidence in a court of law. Specimens collected from home are not strongly supported in as evidence in a court of law.

Enclose pictures and place fingerprints in appropriate areas provided:

Parent's fingerprint	Alleged parent's fingerprint	Child's fingerprint

Identification: (Optional)
Mother's Name: ______________________
ID type: __________ ID# __________

Father's Name: ______________________
ID type: __________ ID# __________

Child's Name: ______________________
ID type: __________ ID# __________

Packaging, Witnesses and Specimen Collection:
I certify that I have collected samples from the persons whose pictures and names appear on this form and mailed them into GeneTree Inc. for analysis.

Name: ______________ Initials: ____ Date: ______
Location: ______________________
Phone: ( ) ______________
Packaged by: ______________ Initials: __________
Method of shipment: ______________________

Tracking #: ______________ Date: __________

Witnessed by: ______________________
Initials: ___ Date: ______
Witnessed by: ______________________
Initials: ___ Date: ______

Therefore, some laboratories may ask whether the alleged father being tested has any close relatives that could also be the father of the child. This alerts the laboratory to watch for genetic patterns that may be similar, but not identical, to the child. Irrespective of whether the question regarding close relatives is asked, most laboratories examine a sufficient number of genetic markers to exclude non-fathers (even those individuals that are closely related to the biological father). But it is recommended that both alleged fathers be tested, whether or not they are related.

The specimens analyzed by most paternity testing laboratories are collected either by blood or buccal swabs. Years ago, a tube of blood was drawn by venipuncture to obtain sufficient DNA for analysis. The methods now used for paternity testing are extremely sensitive and the analysis requires very small amounts of DNA from each person. Laboratories that analyze blood usually collect only a few drops by pricking the end of a finger. The blood may be spotted to a solid medium such as a piece of paper or cloth for transport to the laboratory. Blood collected from infants and young children is usually obtained by pricking the heel. The most common procedure for collecting DNA at present is to rub the inside of the cheeks with buccal swabs (Fig. 2-2) that are specially designed to collect cheek (buccal) cells. Collecting buccal cells is non-invasive, and involves very little discomfort. The swab can be forwarded to the laboratory for analysis or it may be blotted to a filter paper to transfer buccal cells to a paper for transport and storage.

In a typical 'legally admissible' paternity test, each person's specimen(s) are sealed in an envelope or bag and the group of specimens collected for a case are sealed in an evidence bag for transport to the laboratory. The sealing of specimens is part of the Chain of Custody procedures. Upon receipt, the laboratory performing the analysis will examine the specimens, note their condition and packaging upon arrival, and determine whether the specimens have arrived to the laboratory without evidence of tampering.

After completing the analysis, the results of a legally admissible test are usually returned as a formal, notarized report signed by laboratory director and/or DNA analyst. The report includes the names of the parties tested, the race of the mother and alleged father, the date of specimen collection, the type of analysis was performed and the results (Figure 2-3). As noted above, the report also contains a statement indicating whether the alleged father is excluded or not excluded. If the alleged father is not excluded a statement or statements concerning statistical certainty are also provided.

The paternity test process is understood in detail when one understands the basic concepts of heredity.

Figure 2-2

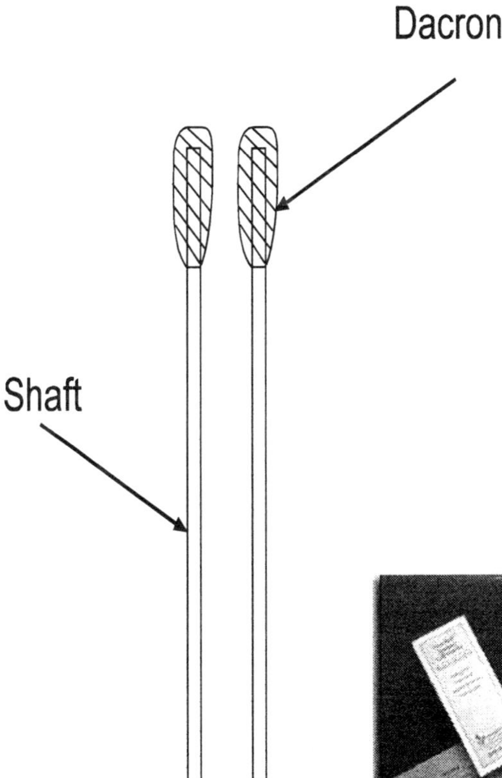

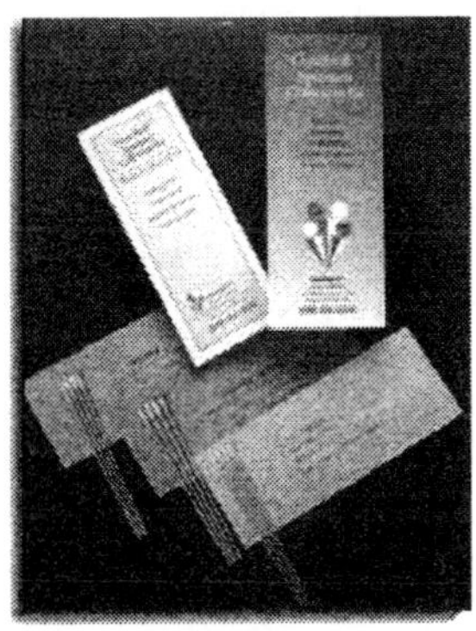

Figure 2-3

GeneTree Inc.
3150 Almaden Expressway, #203
San Jose, CA 95118
http://www.genetree.com/
Toll free: (888) 404-GENE
Phone: (408) 723-2670
Fax: (408) 723-2671

Paternity Analysis Report

(DTL-00001X)

Samples Tested (04/25/00):
Specimen Sample #1: John Doe-Alleged Father
Specimen Sample #2: John Doe Jr.-Child
Specimen Sample #3: Jane Doe-Mother

Services Requested: ☒DNA Profile/Relatedness ☐Forensic ☐Clinical ☐Diagnostic
Case type: ☒Personal ☐Civil ☐Criminal ☐Other:________________

Explanation of DNA Profile:
Four discrete locations on several different chromosomes were used to construct the DNA profiles for each individual tested. Table 1 shows John Doe-Alleged Father to have identification markers 8.24 and 4.32 at the D4S163 STR chromosome location, and John Doe Jr.-Child to have identification markers 7.54 and 4.32 at the same STR chromosome location. This suggests that identification marker 4.32 was passed from the alleged father to the child, and identification marker 7.54 was passed from the biological mother to John Doe Jr.-Child. Results from other identification markers are also listed in Table 1. Each marker has an established probability of occurrence in the general population and among different ethnic groups.

DNA Profiles:

Table 1: The DNA profiles of John Doe-Alleged Father, John Doe Jr.-Child, and Jane Doe-Mother.

Samples:	D2S44	D4S163	D7S467	D10S28
John Doe	3.32	8.24,4.32	5.36,3.85	2.00,1.51
John Doe Jr.	3.32,1.57	7.54,4.32	4.45,3.85	4.70,2.00
Jane Doe	1.57, 2.10	3.27, 7.54	4.45, 2.61	4.70, 12.31

Probability of paternity: >99%

Conclusion:
DNA profiling was performed by standard methods, and has been completed on samples in the name of John Doe-Alleged Father, John Doe Jr.-Child, and Jane Doe-Mother. Based on the observed scientific evidence, it is concluded, for all practical purposes, in reference to the submitted samples, that there is >99% probability that John Doe is the biological father of John Doe Jr.-Child. Therefore, John Doe-Alleged Father is **included** as the possible biological father of John Doe Jr.-Child.

Sample Type: ☐Blood ☐Bone ☒Saliva-Buccal Swab ☐Buccal Brush ☐Hair ☐Semen ☐Tissue ☐Tooth pulps

____
Senior Clinical Scientist

Chapter 3

THE HISTORY THAT LED TO UNDERSTANDING HEREDITY

To understand family genetic relationship and paternity testing one needs to be familiar with basics of biological inheritance. Mythology reflects humans' early concerns about biological inheritance and proof of it since ancient times. In Greek mythology Zeus, the Ruler of Gods on the Olympus Mountains visited Alcmena when her husband, Amphitryon, was away at war with the Taphians. Heracles was conceived during this visitation, and when nine months passed Zeus boasted that on that day a son would be born of his line who would rule over the land of Greece. However, his wife, Hera, was naturally angered at her husband's continuous philandering and decided to thwart him. She sent Eleithyia, goddess of childbirth, to delay Heracle's birth while she ensured that Eurystheus, who was also of Zeus' line through Perseus, grew up to be King of Tirynus and Mycenae.

Alcmena was in pain from the delay to the birth. To relieve her, her servant Galanthis rushed in and startled the goddess Eleithyia, whose attention lapsed, allowing the birth to take place. Subsequently, Alcmena bore twins, the half-divine Heracles and the all-human Iphicles.

The proof of the divine lineage of Heracles was demonstrated soon when, still in the cradle, the baby killed two snakes that were sent by Hera. People were astonished by this miracle, accepted his lineage to the Gods on Olympus and started worshipping the young boy. A miracle like this was enough to prove a paternity relationship to the Gods on Olympus.

The notion of inheritance was present throughout the ages and finally scientifically implemented in the theory of evolution. While developing his theory of evolution by natural selection, the British scientist, Charles Darwin, was unaware of the molecular basis for evolutionary change and inheritance. Around the same time that Darwin was making his landmark observations that led to his publication, *The Origin of Species*, the Augustinian monk Gregor Johann Mendel was performing experiments with garden peas that dramatically influenced the field of biology (Fig. 3-1). Mendel's interpretation of his experimental results served as the foundation for the discipline known as genetics. Genetics is the discipline of biology concerned with the study of heredity and variation.

Figure 3-1

In 1866, Mendel published a classic paper in which he described phenomena in garden peas that laid down the principles for genetic inheritance in living organisms. Our understanding of genetics in higher organisms such as humans has advanced considerably since Mendel's work. This has occurred, in part, because of significant advances in our understanding of the molecular biology of living cells. The molecular biology revolution that led to our current understanding of genetics was greatly accelerated in 1953 by James Watson and Francis Crick, who revealed the structure of DNA as a double-helical molecule.

Although Mendel's landmark principles on inheritance continue to form the basis for our current understanding of genetics in living organisms, the knowledge of genes and chromosome structure, and how gametes form by meiosis, made up the next steps required in understanding inheritance.

INHERITED TRAITS

The experimental model for Mendel's work involved performing cross-pollination experiments with a strain of garden peas called *Pisum sativum*. Mendel was interested in studying the inheritance of a number of different characters, or heritable features, of *Pisum*. He considered seven different characteristics including flower color, flower position, seed color, seed shape, pod color, pod shape, and stem length. Variations of a given character are known as **traits**. For example, when studying flower color as a characteristic, Mendel traced the inheritance of two traits for flower color, purple and white. Many of the basic genetic principles established by Mendel arose from his observations of the results produced by a simple cross-pollination experiment called a monohybrid cross. In a monohybrid cross, individuals with one pair of contrasting traits for a given character are mated. For example, a plant with purple flowers is mated with a plant that has white flowers.

Mendel proposed several basic principles to explain inheritance. One of his first principles was that characters are determined by what he described as "heritable factors" or "units of inheritance." The factors that Mendel described are what we now call genes. Mendel noted that alternative forms of a gene,

Mendel's laws are used to create four rules for parentage testing:

1. The child can not have a genetic marker that is absent in both parents.
2. The child must inherit a pair of genetic markers, one from each parent.
3. The child can not have a pair of identical genetic markers unless both parents have the marker.
4. The child must have the genetic marker that is present as an identical pair independently in both parents.

known as alleles, are responsible for variations in inherited characters. For example, in *Pisum sativum* there are two alleles for flower color, a white flower allele and a purple flower allele.

Mendel also proposed that certain alleles, called **dominant** alleles, always expressed themselves in the appearance of an organism and that the expression or appearance of other alleles, called **recessive** alleles, were sometimes hidden or masked. Modern-day geneticists frequently use capital letters to indicate dominant alleles and lower case letters to indicate recessive alleles.

The appearance of a particular trait is referred to as the phenotype of an organism while the genetic composition of an organism is known as the organism's genotype. The genotype determines the phenotype.

Mendel established that an organism typically inherits two alleles for a given character, one from each parent. Using modern genetic terminology, we say that an organism that contains a pair of the same alleles is homozygous for a particular trait while an organism that contains two different alleles for a trait is heterozygous for that trait.

Mendel also postulated that when an organism contains a pair of units, the units separate, or segregate, during gamete formation so that each individual gamete produced receives only one unit of the pair. This postulate became known as Mendel's law of segregation of alleles. This principle is very important in understanding the outcome of paternity tests and will be discussed in following pages of this book.

Another example of a monohybrid cross-performed by Mendel involved the inheritance of color in pea flowers. Mendel began by crossing true-breeding homozygous parents called the ***P*** generation. He crossed a plant with purple flowers with a plant showing white flowers. The first offspring produced from such a cross, called the F_1 generation, all showed purple flowers. The results of this cross were explained when Mendel determined that the F_1 plants were heterozygotes and that these plants showed purple flowers because the purple flower allele was dominant and the white flower allele was recessive. Self-pollination of F_1 plants produced a second generation of plants called the F_2 generation, in which the phenotypic ratio of purple-flowered plants to white- flowered plants was approximately 3:1. As he performed monohybrid crosses for several other characters, Mendel discovered that this ratio was characteristic of a cross between heterozygotes.

Mendel's law of segregation of alleles is very important when scientists are concerned with physical characteristics in humans and they analyze heredity lineage in our family. It is noteworthy that certain characteristics (traits) are determined by one factor or a gene, whereas others are more complicated and governed by multiple factors.

One of the first diseases explained by Mendel's genetic factors was brachydactyly, named for the shortened fingers of affected individuals. In brachydactyly one allele (*B*) can overide the action of its partner allele *b* (Figure 3-2). Thus a man with normal fingers (*bb*) and a brachydactylous woman (*Bb*) will have a ratio of 1:1 of brachdactylous and normal children according to Menedel's law discussed above. This means that there is a 50% chance for each child of the couple to receive the dominant brachydactyly allele and have

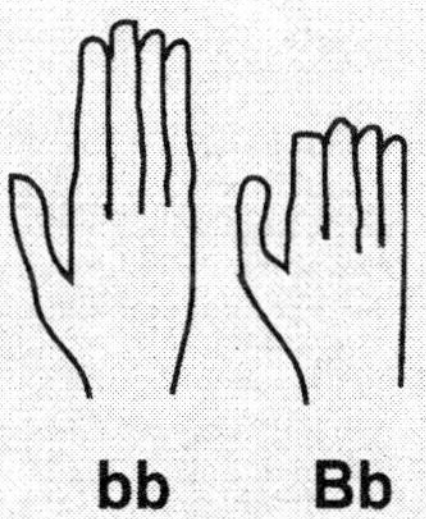

Figure 3-2

Table 3-1

	Father (Genotype Bb)		
Mother (Genotype bb)	**Gametes**	**B**	**b**
	b	Bb	bb
	b	Bb	bb

PUNNETT SQUARE

The genotypes and phenotypes that may result from a genetic cross can be predicted by using a Punnett square. Mendel used dihybrid (two traits being analyzed) crosses to explain his law of independent assortment, in which he postulated that alleles for different characters (for example, pea color and pea shape) segregate into gametes independently of each other. A dihybrid cross between F_1 heterozygotes reveals a phenotypic ratio of approximately 9:3:3:1 in the F_2 generation (see Table 3-2).

The 3:1 ratio predicted for Mendel's monohybrid cross and the 9:3:3:1 phenotypic ratio predicted for a dihybrid cross are hypothetical expected ratios. When performing a real genetic cross based on Mendelian principles, however, such a cross is subject to random changes and experimental errors that produce chance deviation in the actual phenotypic ratios that one may observe. To accurately evaluate genetic inheritance, it is essential that observed deviation in a phenotypic ratio be determined and compared with predicted ratios.

Table 3-2 Symbols: D = tall, W - round, d = dwarf, w = wrinkled

		Plant #1			
		DW	**Dw**	**dW**	**dw**
Plant #2	**DW**	DDWW tall, round	DDWw tall, round	DdWW tall, round	DdWw tall, round
	Dw	DDWw tall, round	DDww tall, wrinkled	DdWw tall, round	Ddww tall, wrinkled
	dW	DdWW tall, round	DdWw tall, round	ddWW dwarf, round	ddWw dwarf, round
	dw	DdWw tall, round	Ddww tall, wrinkled	ddWw dwarf, round	ddww dwarf, wrinkled

The principles established by Mendel can be easily explained by understanding the chromosomal basis of inheritance. For example, in humans, somatic cells, which make up most the cells in the body, contain the diploid number (2n) of chromosomes. These cells have 46 chromosomes organized as 23 pairs of homologous chromosomes

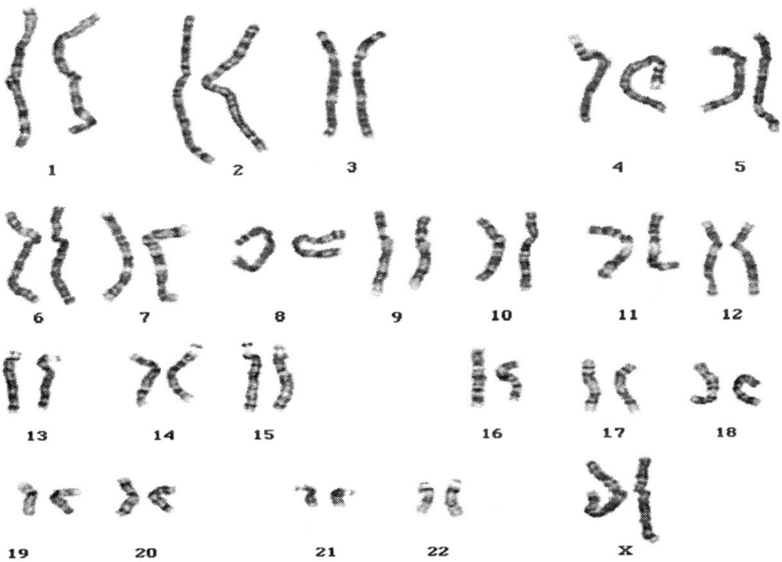

Figure 3-3

(homologues; see Figure 3-3). The molecular nature of chromosomes will be explained in the next chapter. Chromosomes 1 through 22 are called autosomes and the twenty-third pair of chromosomes is called sex chromosomes (X and Y). During meiosis, gamete formation leads to the formation of sex cells (non-somatic cells) that contain a single set of 23 chromosomes - the haploid number (n) of chromosomes. Because chromosomes are present as pairs in human cells, genes located on each chromosome are also typically present as pairs. Mendel's law of segregation of alleles is explained by the separation of each pair of homologous chromosomes that occurs during meiosis, resulting in each individual gamete (sex cells) receiving only one copy of each chromosome. This law applies to, and is explained by, the separation of unlinked genes because these genes are located on different chromosomes. As a result, during meiosis the chromosomes align at the metaphase plate in the center of the cell, and separate in a random, independent fashion, dividing the chromosome pairs into two different sex cells.

An expansion in our knowledge of Mendelian genetics has led to a detailed understanding of different types of inheritance events such as co-dominance, incomplete dominance, linked inheritance, inheritance of multiple alleles, lethal mutations, and the genetic transmission of a number of different genetic disorders. For many reasons, Mendel's pea plants served as an ideal model organism for him to study. In fact, pea plants allowed Mendel to succeed where others failed. Peas are easy to grow, they cross-fertilize and self-fertilize easily, and they mature quickly. More recently, a number of other organisms have served similar roles as model organisms for scientists who study genetics. Some of these include the flowering plant *Arabidopsis thaliana*, a common inhabitant of home aquariums called the striped zebra fish (Brachydanio rerio), many species of mice, and the common fruit fly, *Drosophila melanogaster*.

Chapter 4

UNDERSTANDING THE HUMAN GENOME

DNA (deoxyribonucleic acid) is most commonly recognized as two paired chains of chemical bases, spiraled into what is commonly known as the double helix. This double helix gives the molecular structure an appearance of a spiral staircase. Each chain, or strand, of DNA is made up of a sequence of bases, or chemical building blocks. There are four different kinds of bases in DNA, and this is what makes it simple to understand its basic function and structure. The order in which the bases occur determines the information stored in the region of DNA being looked at.

Each base can be thought of as a specific letter of the alphabet, combining to form words and sentences. Together, DNA works with protein and RNA similarly to how a computer works with programs and data. DNA has a code that is continuously active, having instructions and commands such as "if-then", "go to", and "stop" statements. It is involved with regulating protein levels, maintaining quality control and providing a database for making proteins. The best analogy to use when describing DNA and its function in cells is to look at a cell as a little factory.

The two uprights of the DNA ladder are a structural backbone, supporting the rungs of the ladder. These are the informational parts of the DNA molecule. Each rung of the ladder is made up of two bases that are paired together. This is what makes the steps of the spiral staircase. The two-paired bases are called a base pair. The length of any DNA fragment is measured in base pairs (bp), similarly to how we measure length in inches or meters.

Each base has a discrete identity; the sequence of the identities is read by machinery inside the cell. Genes, which are made up of large sequences of DNA (thousands of base pairs), can be looked at like instructions telling the cell how much protein to make, when it should be made, and the sequence that can be used to make it. The information in the DNA is like a library. In the library you will find books, and they can be read and re-read many times, but they are never used up or given away. They are retained for further use. Similarly, the information in each gene (see below) is read, perhaps millions of times in the life of an organism, but the DNA itself is never used up.

GENES, CHROMOSOMES, AND PATERNITY

A gene is a segment or region of DNA that encodes specific instructions, which allow a cell to produce a specific product. This product is typically a protein, such as an enzyme. There are many different types of proteins. Proteins are used to support the cell structure, break down chemicals, build new chemicals, transport items, and regulate production. Every human being has about 100,000 putative genes (a controversy exists over this number at the moment of this writing) that produce proteins. Many of these genes are always identical from person to person, but others show variation in different people. They are what are used to determine hair color, eye color, sex, personality, and many other traits that in combination make everyone a unique entity. Some genes are involved in our growth and development. These genes act as tiny switches that direct the specific sequence of events that are necessary to create a human being. They affect every part of our physical and biochemical systems, acting in a cascade of events turning on and off the expression, or production, of key proteins that are involved in the different steps of development.

The key term in growth and development is differentiation. Differentiation involves the act of a cell changing from one type of cell, when dividing through mitosis, into two different types of cells. The most common cells used to study differentiation are the stem cells. These are considered to be the 'mother cells', and are thought to be capable of differentiating into any type of cell. It is important to understand the value of differentiation when learning about genetics and human development. Everyone starts life as one single cell, which divides several times before any differentiation takes place. Gene products are produced to carry out the genes' function. To make products from a gene, the information in the DNA is first copied, base for base, into a similar kind of information, called a transcript, or RNA (Figure 4-1).

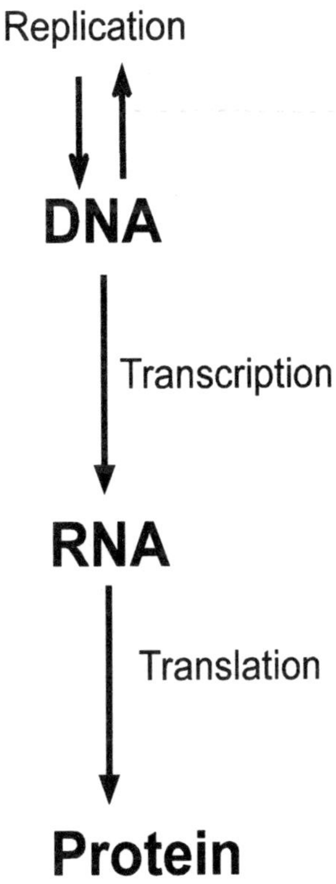

Figure 4-1

The RNA copy of the gene sequence acts as a messenger, taking information from the nucleus (where the DNA is found in its chromosomal form) and transporting it into the cytoplasm of the cell (where the machinery for making gene products is found). Once in the cytoplasm, the messenger RNA is translated using a different chemical language into the product of the gene, a protein. The sequence of the protein is defined by the original sequence of the DNA bases found in the gene.

Mutations are changes in the DNA information. Mutations can be inherited from a parent, or may be acquired during the life of an organism (due to random events or to environmental influences). Mutations can cause protein products to be altered, missing from the cell, or present in incorrect amounts. Mutations are important to understand because they are what cause many genetic diseases.

Single nucleotide changes in the DNA sequence (Figure 4-2), which do not cause any deleterious effects are termed Single Nucleotide Polymorphisms (SNPs). They are used as genetic markers in large population or distant family relationship studies.

Genes make up only a subset of the entire amount of DNA in a cell. Spacer DNA, which conveys no protein-coding or regulatory information, makes up the bulk of the DNA in our cells. Active research in the last 20 years has identified

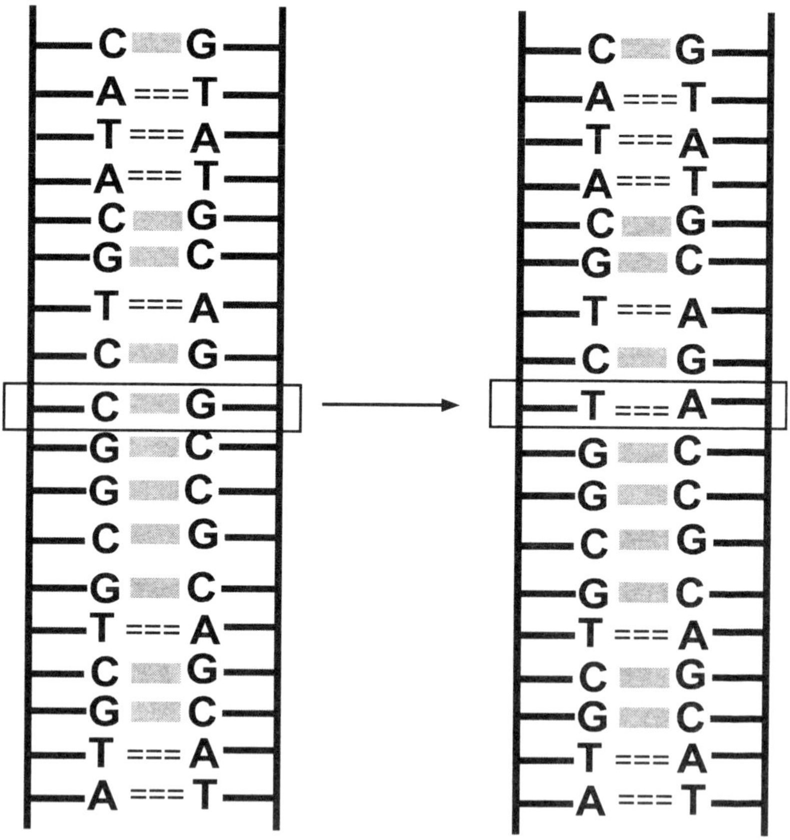

Figure 4-2

signals (specific sequences of bases) that delimit the beginning and ending of genes. These signaling areas can be thought of as regulatory elements. They are what is used to control the production of protein, and work on a biofeedback system. They are usually located near the beginning of a sequence that is used to code for a protein. Through protein-DNA interactions they are used to turn on and off the production of proteins.

CHROMOSOMES

Genes are found packaged into the 46 chromosomes that are located in the nucleus of most of our cells (Fig. 4-3). The cells

Figure 4-3

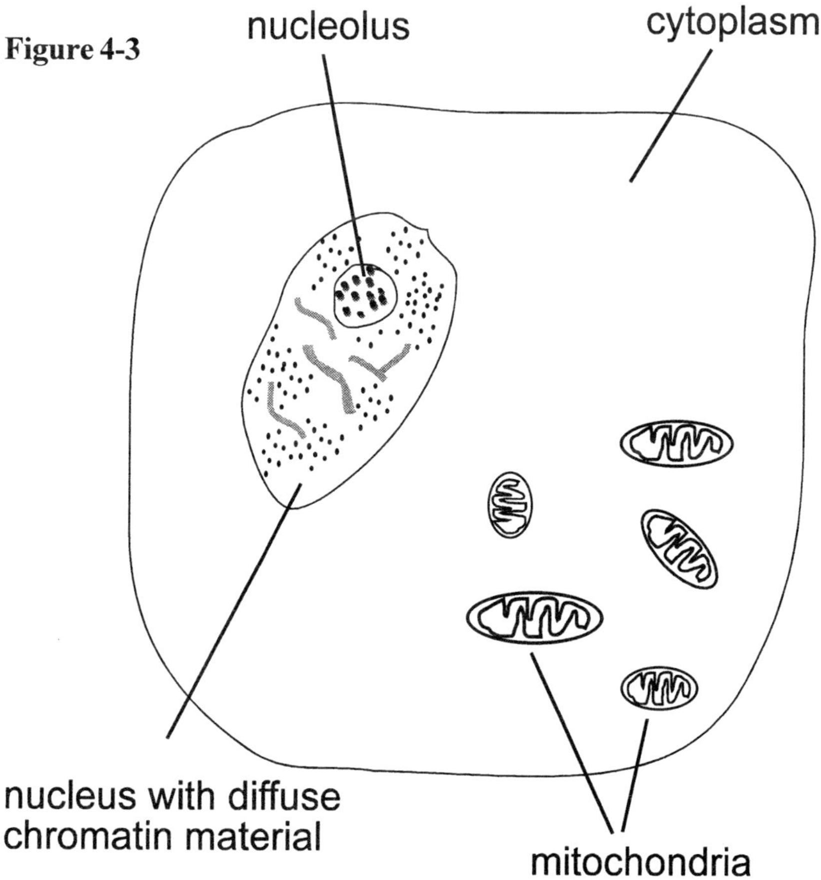

are called diploid (2n, which indicates 2 sets of chromosomes). A chromosome is a complex of a DNA molecule and proteins packed tightly together. There are two copies of each chromosome in the human cell, one copy from the paternal (father) side and the other from the maternal (mother) side.

Sperm and egg cells are different from other cells because they have only 23 chromosomes, i.e. they are haploid (n) with only one set of chromosomes. In males the sex chromosomes are different. Males have one X chromosome and one Y chromosome. Females have two X-chromosomes. Egg formation (oogenesis) and sperm formation (spermatogenesis) occurs through a complicated mechanism of cell divisions known as meiosis. During this process diploid cells (2n) give rise to haploid cells (n). An important phenomenon happening during this process is the shuffling or exchange of DNA segments of non-sister chromatids, called crossing-over.

Human spermatozoa (sperm cells) travel a tortuous path toward the egg after being deposited in the vagina. Spermatozoa cross through the uterix cervix and cavity, find the opening of the fallopian tube, enter the tube, and after an additional way, find one egg, and fertilize it. When a single sperm and egg come together — when pregnancy begins — they form their own new cell with 46 chromosomes. The human being that results is genetically unique, with a blueprint half from each parent. The sperm is what determines the sex of the child. Each sperm will have either an X chromosome (female) or a Y chromosome (male).

GENOMIC DNA IN PATERNITY TESTING

The information passed from our natural parents is encoded in DNA, which is contained in the cell nucleus of most cells in our body. Cells are small structural and functional units of a multicellular organism. Cells are specialized but they share similarity in regard to their structure and intracellular functioning.

Human cells contain membranes that form functional compartments within the cell. Compartmentalization is the basis of evolution of organisms on this planet. All of the genetic material of a cell nucleus is called a genome, and the DNA isolated from the nucleus is termed genomic DNA. This DNA is used in most paternity testing procedures. Since half this DNA is given from the father and half from the mother, it is easy to use genomic DNA to determine paternity (or maternity) issues.

The genome is spread out among multiple chromosomes, each of which is a single linear DNA molecule. DNA is tightly associated with proteins called histones that form a complex around which approximately 200 bp of DNA are wound (Figure 4-4).

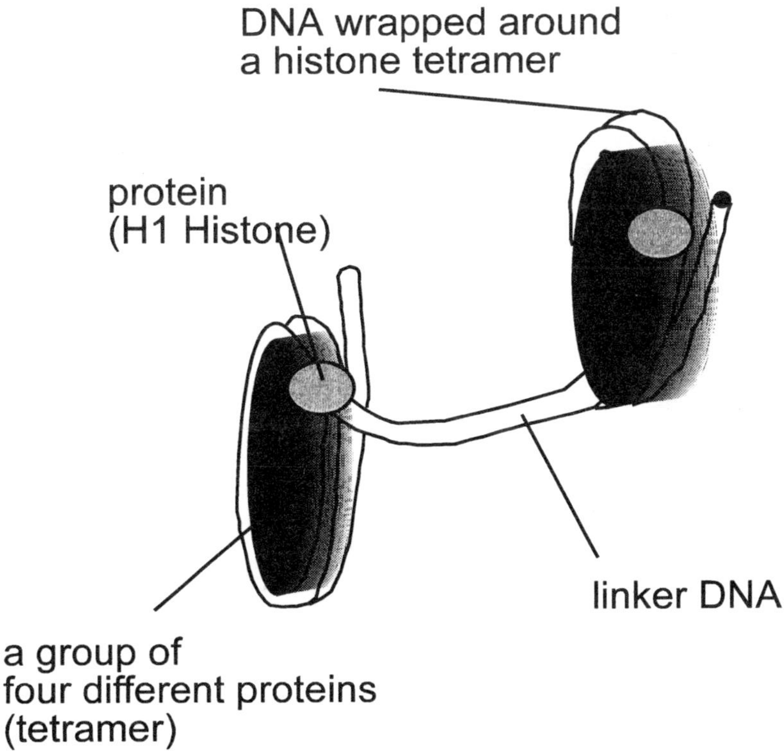

Figure 4-4

Each molecule of DNA incorporates hundreds of these nucleosome structures in a continuous chain.

The objective of DNA isolation in a paternity testing protocol (and any other protocol involving the analysis of DNA, such as genetic testing, research, etc.) is obtain useful and "clean" DNA that is free from contaminating molecules. This means that cells have to be separated from their environment, and all the compartments encompassing the genomic DNA have to be broken and separated from the DNA molecules. In addition, undesired additives used to facilitate these separations may also need to be removed. This involves procedures that used to be very tedious, called DNA extraction and purification. However, with the availability of DNA isolation kits from commercial vendors and standardized procedures, this step can now be performed easily and accurately.

MITOCHONDRIAL DNA

Cells in our bodies possess energy generators called mitochondria. These cell organelles are another proof for the importance of compartmentalization of life evolution on this planet. Mitochondria have enzyme systems and other specialized proteins, which extract energy from the breakdown products of the foods we consume. These enzymes are encapsulated by a double membrane structure within the cell. Mitochondria are most numerous in very active cells such as muscle, heart, kidney, nerve, and secretory tissue.

An amazing feature of mitochondria is the presence of DNA in them, which is completely separated from the genome contained in the nucleus of the human cell. This DNA is called mitochondrial DNA (mtDNA) or mitochondrial genome and was first sequenced by biochemists at Cambridge University in London, England in 1981. The leading author of the scientific article reporting the mitochondrial sequence was S. Anderson and the sequence is known as the Anderson sequence. It serves as a reference to which subsequent analyses are commonly compared.

An important difference between nuclear and mitochondrial DNA lies in the pattern of inheritance. Mitochondria (and therefore mitochondrial DNA), come from the egg of an individual's mother with little or no contribution from the father. The reason for this is the loss of much cytoplasm during spermatogenesis. The remainder is preserved in the head of the sperm cell, which contains no mitochondria.

Mitochondria are found only in the upper part of the sperm tail. They provide the energy for the movement of sperm cells. The sperm's tail, mid-piece and nuclear envelope disintegrates

Table 4-1

CHARACTERISTICS	NUCLEAR DNA	MITOCHONDRIAL DNA
Size	~ 3 billion bp	16,569 bp
Copies per cell	2	Can be > 1,000
Structure	Linear, packaged in chromosomes	Circular
Inherited from	Father and Mother	Mother
Generational Recombination	Yes	No
Mutation Rate	Low	At least 5-10 times higher than nuclear DNA
Sequence	Sequence completed in 2000	Described in 1981

after the sperm and the egg fuse. In this way mitochondria in the newly formed cell come only from the mother.

In contrast, nuclear DNA is a combination of DNAs from both parents. Nuclear DNA goes through recombination or exchange of inheritance material between the paternal and the maternal sets. Mitochondrial DNA represents only the maternal ancestry of an individual.

Unlike the nuclear DNA, mitochondrial DNA is found in high copy number, i.e. 1,000-10,000 copies of the mitochondrial genome per cell. Therefore, mtDNA is more abundant than nuclear DNA in a sample of extracted DNA from cells. This is important when dealing with buccal swabs in determining family relationships or in working with highly degraded biological samples. Additionally, mtDNA is enveloped in a protective double membrane, which is resistant to environmental stress. The use of mtDNA in establishing family relationships will be discussed in a subsequent chapter.

BIOLOGICAL SOURCES FOR DNA

We have a handy, if unsavory, example of the knots DNA testing can unravel in the civil lawsuit filed against the estate of Larry Hillblom, one of the founding partners of DHL Worldwide Express.

In the late 1960's, three young men realized they could legally duck import taxes on cargo by flying the bills of lading in ahead of the ship carrying the cargo. In 1969, they began ferrying the documents between San Francisco and Honolulu. The company grew quickly, virtually dominating the Pacific Rim, and making it's founders very wealthy.

Larry Hillblom loved to fly, especially relishing the flights he made in his vintage aircraft. His passion finally got the better of him, however, when his plane crashed into the waters off Saipan, where he lived part time.

His estate was valued at approximately a half billion dollars, but his will was contested by four unexpected claimants, who were eventually able to prove that they were indeed his unacknowledged offspring, by four different young women. Initially, it was their word against the Hillblom's estate's lawyers. They thought they first had supporting evidence when the child-claimant lawyers found a San Francisco hospital that in 1993 had removed and preserved a facial mole of Hillblom's. Interestingly, the mole was later found not to be Hillblom's, and the fallback position was to perform a siblingship analysis to support the claim that all the children came from the same father. This test showed just that, but it was not enough to convince the Hillblom estate lawyers that the father was Hillblom himself. Therefore, a grandparentage analysis was performed using a drop of blood purchased for $1,000,000 from Hillblom's mother. This proved the claim was accurate, and also demonstrates the flexibility of DNA testing to determine biological relationships. Furthermore, this illustrates that where there is a 'will' there is a way -to use DNA.

Biological material used in paternity testing is usually either blood cells or cheek cells, but as read earlier, almost any tissue can be used. Cheek cells are easily collected, and involve scraping the inside of the cheek with a ‘Q-tip-like’ swab. The quantity of the collected cells is important for successful DNA-based testing. Many labs prefer to use blood for DNA based testing, but it is becoming a routine procedure to use cheek cells because of the well-developed protocols for genomic DNA extraction. Some laboratories prefer blood because such samples yield larger quantities of DNA. The fact that DNA based testing methods are becoming more and more sensitive has led to a dramatic decrease in the amount of required starting material.

Genomic DNA is extractable from either fresh or frozen blood samples. Although DNA can also be isolated from clotted blood, homogenization of the clot is required, and yields can be lower. It is, therefore, preferable to collect blood samples in the presence of an anticoagulant such as heparin or EDTA. There are various protocols in clinical laboratories for blood collection, but heparin and EDTA are the most commonly used anticoagulants. It has been determined that the use of citrate dextrose produces higher yields of DNA from blood stored at room temperature. Heparin has been reported to inhibit DNA binding enzymes used in subsequent DNA manipulations necessary for determining paternity. There are many kits available to purify DNA from samples such as blood and cheek cells.

Unlike the red blood cells, leukocytes have nuclei. Leukocytes make up only a small amount of the total number of blood cells found in a 5-ml sample. However, they are easily separated from plasma and red blood cells by centrifugation. The resulting layer of white blood cells, called a buffy coat, is removed and resuspended in an appropriate buffer used for the DNA isolation.

Clearly, cells from the inside of the mouth are routinely used for DNA based paternity testing. Although DNA isolated from cheek cells are not in great quantity or quality, they do give more than enough DNA for reliable analysis.

Isolation of genomic DNA from cheek (and blood) cells involves first breaking the cell membranes. This procedure was changed dramatically when Proteinase K was introduced. Proteinase K is an enzyme, which degrades proteins. Cell membranes are disrupted by a simple incubation at 60°C for several hours in a buffer that contains a detergent such as SDS, Proteinase K, and EDTA. The detergent serves to dissolve the cell membrane and denature proteins. EDTA chelates divalent ions that are required for the activity of the nucleases (proteins that degrade DNA). The basis of this method is that Proteinase K is highly active in this mixture in order to inhibit the activity of DNA-degrading nucleases. The final result is the stripping of DNA from proteins and exposing it to the environment, while at the same time protecting it from degradation. This incubation is followed by purification, usually involving a series of organic extractions, generally employing mixtures of phenol, chloroform, and iso-amyl alcohol. Many paternity testing labs use commercially available kits for DNA purification to ensure consistency of DNA quality and quantity.

The isolated genomic DNA from buccal swab samples is not enough to be visualized by even some of the most sensitive methods. Therefore, amplification of the regions of interest is necessary. The best analogy used to describe the amplification of DNA is to compare it to the process of making copies of paper sheets. The objective of this procedure is to accumulate millions of DNA fragments corresponding to the region of interest and visualize them during analysis as described later in the book. This process is called PCR, and has been associated in popular cases such as the OJ Simpson trial.

Chapter 5

PCR: AMPLIFYING YOUR GENES

The Polymerase Chain Reaction (PCR) has revolutionized the biological sciences by becoming a routine component of nearly every biology laboratory. PCR is used to exponentially amplify target DNA, and can produce millions of copies of a single DNA fragment in a short period of time.

PCR: THE HISTORY

PCR dates back to 1985 when the inventor of PCR, Kary B. Mullis, serendipitously stumbled onto it while working for one of the first bio-tech companies in the US, Cetus Corporation. The idea for PCR struck Mullis while he was pondering on a project involving the sensitivity of a Beta-globulin sensor, a protein found in the blood that serves to transport certain materials. When the idea for PCR came to Mullis, he devoted all of his efforts to making it work. The proof that Mullis was looking for came from the combined efforts of various Cetus scientists in 1985. Mullis received the Nobel Prize in Chemistry for his invention in 1993. PCR has a significant impact on science and has been cited in thousands of publications since, and has over the years became the most commonly used methods in biotechnology. PCR is used in paternity and family relationship extensively, and alternatively to RFLP, which is an older and more accepted method..

DNA - The Template for PCR Amplification

DNA is used as the starting material for PCR in paternity tests. A small quantity of DNA can be duplicated exponentially by combining a temperature cycling condition with a few necessary components. The target DNA is a specific segment of DNA, usually with in a larger piece of DNA or a part of a whole DNA genome. The target is specifically amplified because it is the area of interest to a researcher, either a gene or some functioning unit of DNA that is being studied.

Nucleotides: The Single Units of DNA (A, C, G, and T)

Nucleotides are the building blocks of DNA. Each nucleotide is composed of an alkaline carbon-nitrogen ring structure bonded to a deoxyribose sugar molecule at its 1' end, and is bonded to a phosphate group at the 5' end. The 3' end is used for bonding to the phosphate group of an adjourning nucleotide, forming the structure of a DNA strand.

PCR is a method where a certain region of DNA is copied (amplified), producing at optimum a billion copies that are completely identical to the original.

The nucleotides are simply termed adenine (A), cytosine (C), guanine (G), and thymine (T). The differences between the DNA nucleotides are found in the ring structure. Nucleotides that contain two alkaline C-N ring structures are known as purines (adenine and guanine) and nucleotides that contain one alkaline C-N ring structure are known as pyrimidines (cytosine and thymine).

Free nucleotides are required for DNA replication since they are the building blocks used in extending the DNA chain. They can be thought of as the links that make up a long chain. Others have used musical notes, or the letters of the alphabet as analogies to describe how single nuleotides can be used to construct a multitude of different genes and functional regions. Similarly, musical note can be used to form many different songs or letters can be used to construct many books.

Primers: Giving the PCR Direction

Primers are used to provide the direction to the PCR reaction. A primer is a short sequence of single stranded DNA, called an oligonucleotide. Primers are designed to match the ends of the target DNA required for amplification. These primers become incorporated into the newly synthesized DNA products.

***Taq* DNA Polymerase: The Engineer**

DNA polymerase is a cellular protein that is responsible for copying, or replicating, a DNA template. Through a series of interactions, DNA polymerase basically follows along the DNA template, replicating it by incorporating nucleotides into a separate and new DNA chain, while very accurately mimicking the sequence that it is moving along.

Klenow fragment, the first well-characterized DNA polymerase was commonly used in PCR reactions carried out during the early years. However, Klenow could not sustain the high temperatures required to break the hydrogen bonds. Necessary steps in PCR causing the DNA molecule to split apart into it separate strands. Because of this, Klenow had to be supplied after every temperature cycle, limiting the amount of product and greatly increasing the time involved in PCR.

While in search of a better polymerase than Klenow, *Thermus aquaticus*, a thermophilic, bacterial microorganism that is capable of growing at temperatures 70 to 75 degrees Celsius, was isolated from a hot spring at Yellowstone National Park. The DNA Polymerase native to this species was subsequently thoroughly studied and purified, and then later coined *Taq* DNA Polymerase. Since *Taq* can withstand the high temperatures achieved during the denaturing (strand separating) step in the PCR cycle, it can remain in the reaction. Thus, the step which previously required adding additional Klenow after every heating step had been eliminated, allowing for an easy in automating the cycling procedure.

THE PCR PROCEDURE

The basic components of the PCR reaction are DNA, nucleotides, DNA primers, DNA Polymerase and the right buffer composition. PCR is most typically accomplished in very small volumes, usually measuring 20 to 100 microliters.

PCR amplification involves 3 steps:

1) The 'denaturing' step: The DNA is divide into two separate strands by an increase in temperature that breaks the hydrogen bonds holding the DNA helix together. In a typical PCR reaction, the denaturing step takes place at about 94 degrees Celsius for about 20 seconds. Figure 5-1 shows the first cycle of the PCR reaction.

2) The 'annealing' step: The DNA primers are siting themselves down on the DNA template. The primers bond to the beginning and end of the templates at a temperature corresponding to their calculated melting temperature. This temperature varies from sequence to sequence, but is usually between 45 and 60 degrees Celsius. This temperature is held for approximately 20 seconds each cycle.

3) The 'extension' step: The complimentary DNA strand being synthesized from each primer is extended using the DNA polymerase. Extension occurs at 72 degrees Celsius, and the step is usually performed for 30 seconds each cycle. Figures 5-2 and 5-3 show the second and third cycles of a PCR. The DNA fragments with desired length are marked with arrows in Figure 5-3. Their population will

Figure 5-1

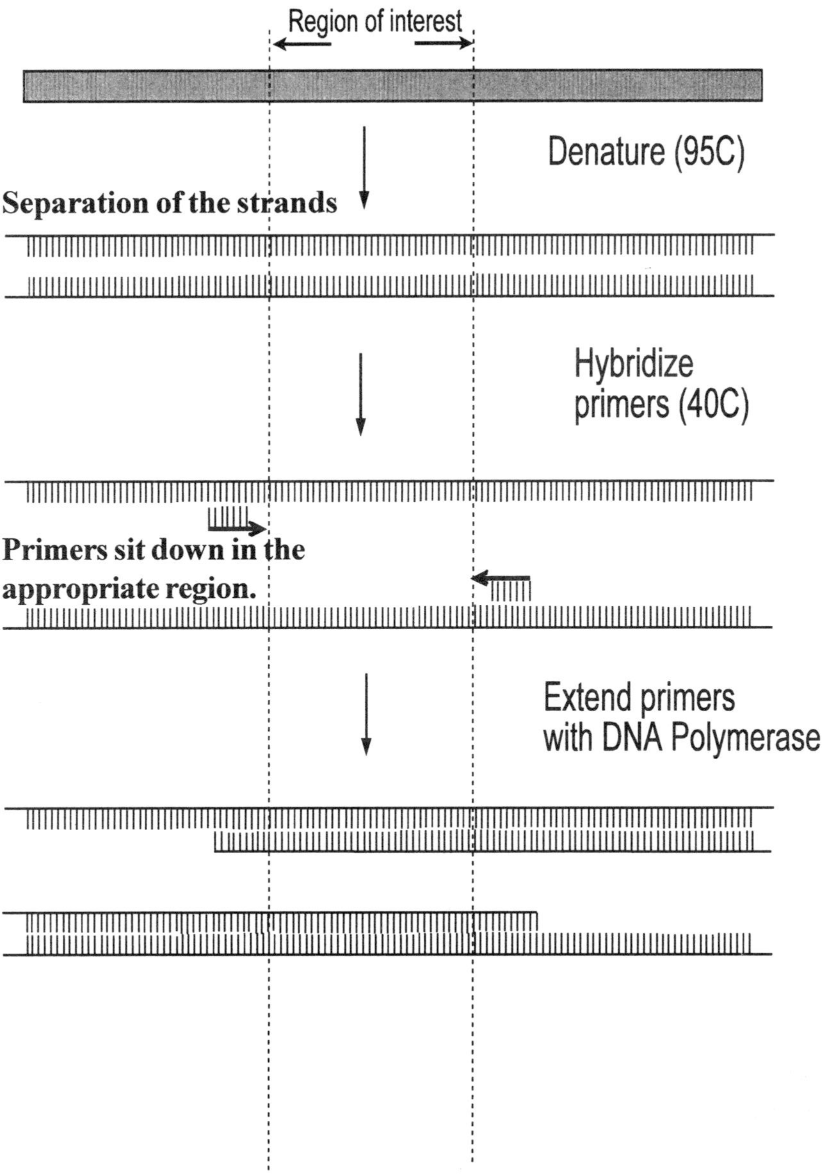
Region of interest
Denature (95C)
Separation of the strands
Hybridize
primers (40C)
Primers sit down in the
appropriate region.
Extend primers
with DNA Polymerase

Figure 5-2

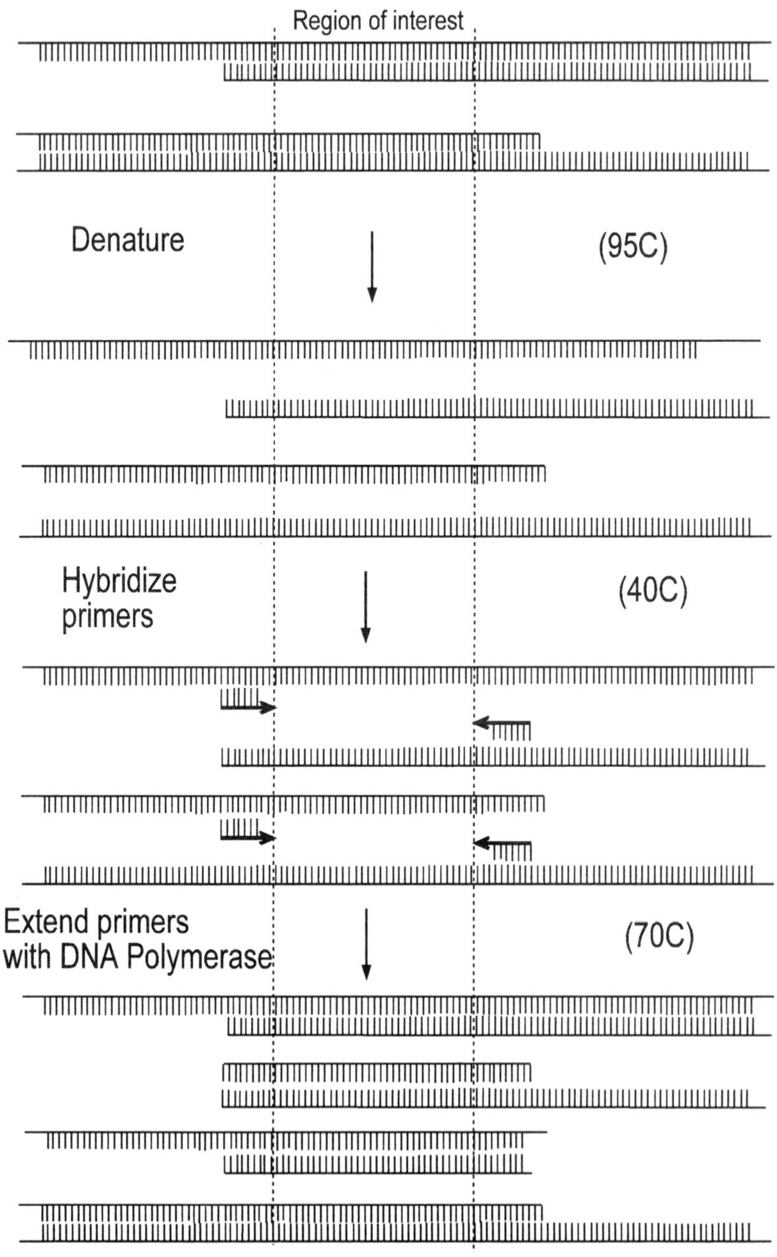
Region of interest
Denature
(95C)
Hybridize
primers
(40C)
Extend primers
with DNA Polymerase
(70C)

Figure 5-3

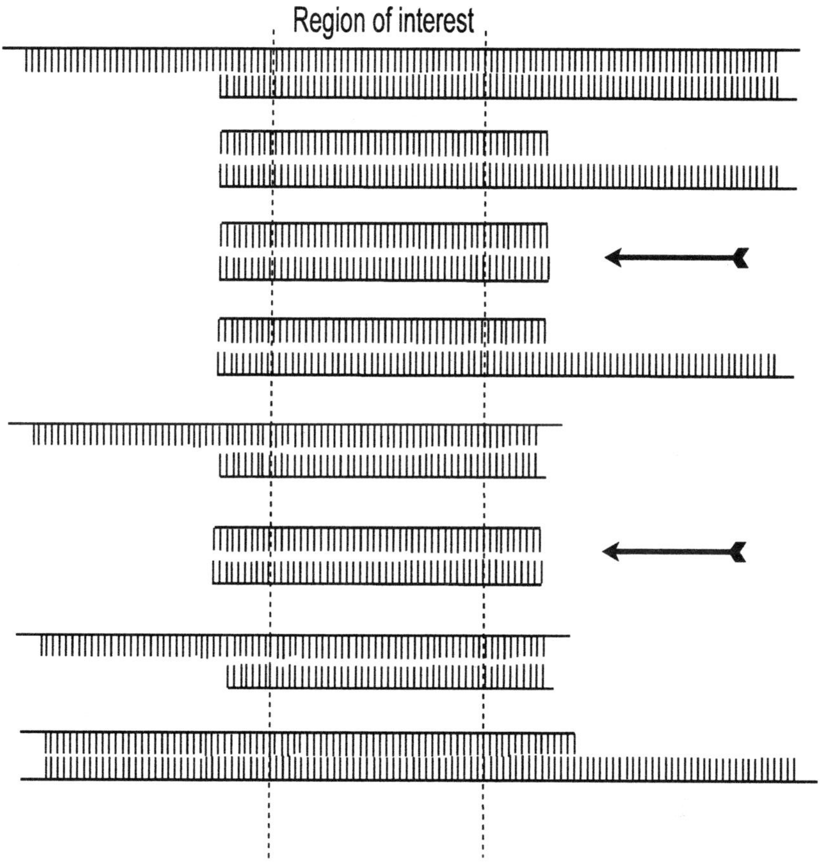

grow exponentially during the following cycles of the chain reaction.

The time involved in carrying out a PCR varies with the amount of desired product and the starting material being used. Optimization is sometimes necessary to get the desired response. Paternity laboratories have established strictly controlled procedures and standardized protocols for carrying the PCR.

NOTES:

Chapter 6

DNA FRAGMENT SEPARATION TECHNIQUES USED IN PATERNITY TESTING

ELECTROPHORESIS

DNA can be either cut into pieces (digested) or specific regions can be amplified by PCR for analysis in paternity testing. These DNA fragments after amplification or digestion then need to be separated from each other and the samples are subsequently compared. Electrophoresis is a technique used in research and clinical laboratories to separate DNA fragments by size. This technique is very powerful as a separation tool, and it is quite sensitive for detecting the differential migration of macromolecules (such as DNA fragments) with only subtle differences in structure or size.

To use electrophoresis, DNA samples must be loaded into wells of an agarose or polyacrylamide gel (Figure 6-1A). These wells act as a starting point for the DNA. An electric field is then applied across the gel, which causes the DNA fragments to move from their origin (the wells) through the gel matrix toward the positive electrode (Figure 6-2B). The gel matrix acts as a sieve through which smaller molecules migrate faster than larger ones. In this way restriction fragments or amplified DNA products of different sizes separate into distinct bands during electrophoresis (Figure 6-3C). The characteristic pattern of the bands produced by each restriction enzyme is made visible by several methods. The DNA can either be previously labeled with the dye, hybridized with other labeled

Figure 6-1

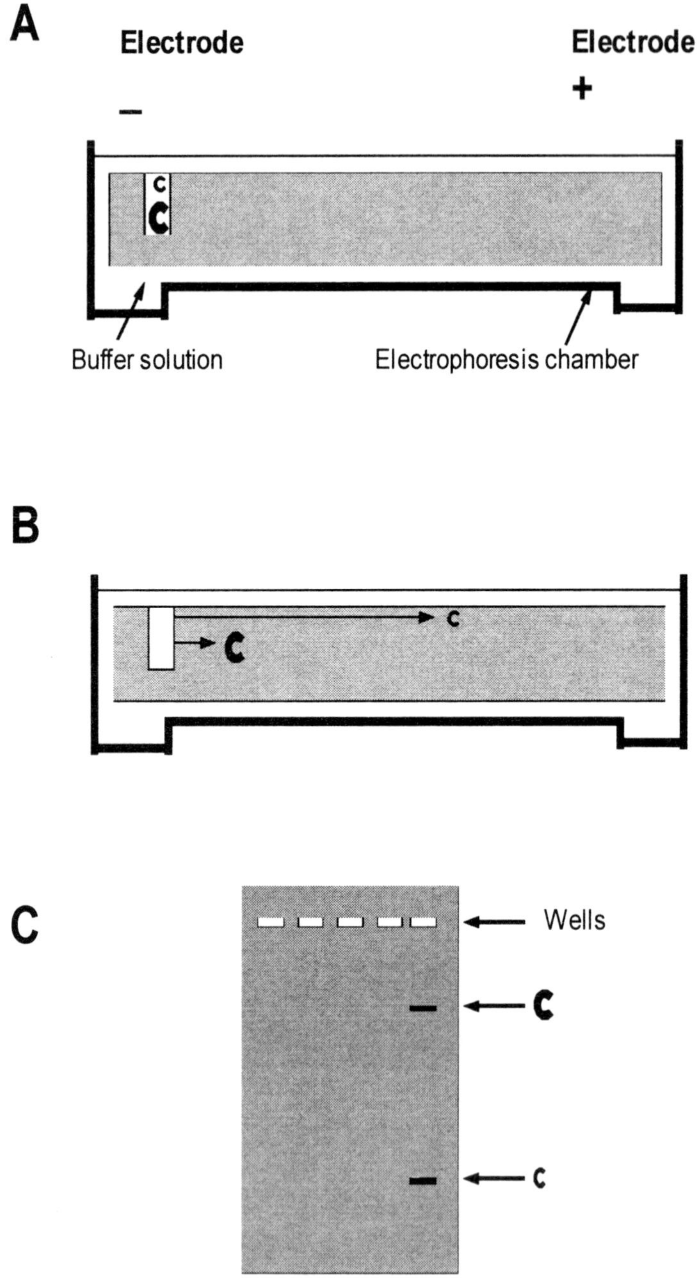

A
Electrode
–
Electrode
+
c
C
Buffer solution
Electrophoresis chamber
B
C
c
C
Wells
C
c

DNA (see Chapter 7), or the dye can be an intercalating chemical that is stained later, after the migration.

There are two main types of gels in common use in DNA laboratories: agarose and polyacrylamide. Agarose is a polysaccharide commercially derived from seaweed. The agarose polymer consists of agarbiose molecules linked together into linear chains. Agarose is supplied as white, nonhydroscopic powder. A gel is prepared by mixing agarose powder with buffer, boiling the mixture, pouring the molten gel into a casting tray, and cooling. During this process the agarose chains shift from existing in solution as random coils to a structure in which the chains are bundled into double helices. The average pore size for agarose gels is typically in the range of 100-300 cubic nm. Agarose gels are used at concentrations near 1% (w/v) for separating DNA fragments in the 1-20 kb (kilo base, 1 kilobase equals 1000 nucleotides) size range. Although agarose gels are easily prepared and inexpensive to use, the only draw back is their limitation in resolving DNA with small changes.

Polyacrylamide gels are prepared from a monomer, acrylamide, and a crosslinker, typically bis-acrylamide. Acrylamide is a potent neurotoxin and is readily absorbed through the skin. Laboratory personnel working with acrylamide should pay great attention when handling this chemical. The concentration of the acrylamide monomer and the ratio of the monomer to crosslinker determine the pore size of the final polyacrylamide gel. Polyacrylamide gels give superior size resolution in the 100-1000 bp size range of DNA fragments.

DNA based paternity testing is performed by determining the size of DNA fragments from particular regions of the genome. Then the sizes found are compared between the different subjects in the analysis. Loading the samples on a gel together with known size markers (DNA fragments of known length that are used as controls) does this. Many kits for paternity testing are commercially available to labs capable of performing these tests. Most kits use DNA that is labeled with fluorescent dyes and the signal read by a detector that

the migration distance of the fragments in the gel, and these distances correspond to the fragments lengths. A standard curve is prepared by plotting the logarithm of the migration distance for each band in the size marker versus its size, which is measured in base pairs. The migration distance of the band of interest in the sample is plotted and the size of the fragment read off the graph. In current, sophisticated and accurate systems, software is used to do this automatically and provide reliable results.

CAPILLARY ELECTROPHORESIS

Capillary electrophoresis (CE) is an alternative method to slab-gel electrophoresis and has the advantages of increased speed and automation. Pouring gels is difficult to do and quite time intensive. Capillary electrophoresis separates molecular species within the lumen of a small capillary filled with an electrolyte. The capillary is immersed in electrolyte-filled reservoirs containing electrodes connected to a high voltage power supply. A sample is introduced at one end of the capillary (the inlet) and DNA fragments are separated as they migrate through the capillary towards the outlet end. As separated fragments migrate through a section at the far end of the capillary, a fluorescent dye detector senses them, and an electronic signal is sent to a recording device.

The injection, separation, and detection are fully automated in capillary electrophoresis methods. Rapid separations (from high voltages) are permitted as a result of efficient heat dissipation in the capillary. In addition, quantitative information is available in a single step in conjunction with computer collection of the data. Very attractive benefits of this system are the minute quantities of sample required and consumed per run and the little waste generated. Because capillary electrophoresis instruments process samples in a sequential rather than a parallel fashion, the results between multiple separations need to be related. Internal standards are usually included in the run. Alternatively, external standards, which are run after a few samples, can be used.

Figure 6-2

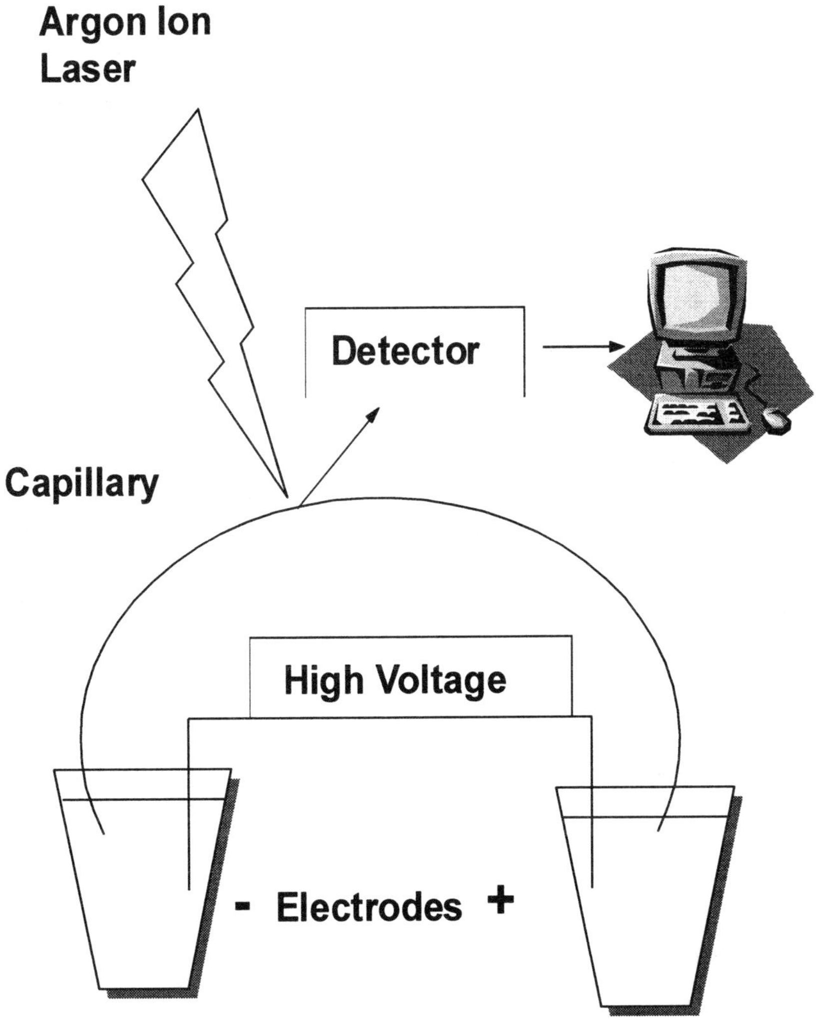
Argon Ion
Laser
Detector
Capillary
High Voltage
- Electrodes +

NOTES:

Chapter 7

DNA PROFILING: A UNIQUE METHOD FOR IDENTIFICATION

One of the first methods used to profile DNA is called restriction enzyme fragment length polymorphism (RFLP) or restriction digest fingerprinting. In humans, RFLP loci with as many as 80 different alleles or variants have been reported. These types of loci are termed variable number of tandem repeats (VNTR), and consist of repeated sets of nucleotide sequences (Figure 7-1). The length of the core repeats can

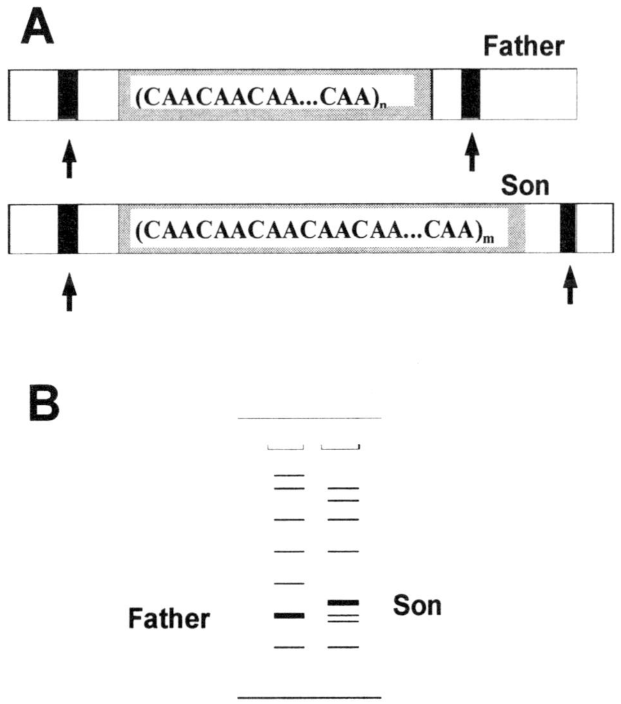

Figure 7-1

vary anywhere from 11 to 60 base pairs. Conserved endonuclease restriction sites, which are short sequences of DNA that can be cut with a particular enzyme, flank the repetitive region. Therefore, once the region is cut out its length can be measured and will be proportional to the number of core repeat units. Each person should have two alleles in these regions. The two alleles found associated with a particular person may either be of the same length or of different lengths. One of these alleles would be passed from a parent to their offspring.

To do RFLP analysis, DNA is first extracted from human body fluid or tissue. Usually more and higher quality DNA is required for RFLP then can be obtained from a buccal swab. Under these circumstances blood is the preferred starting material. After the DNA has been extracted it is digested with a restriction enzyme to cut the DNA into specific fragments. Then it is subjected to gel electrophoresis to separate the fragment sizes out on a gel. The fragments are transferred to a membrane in a method called Southern blotting to fix the DNA on a solid support where it can be accessible to labeled DNA probes (Figure 7-2). The fixed DNA is then probed

Figure 7-2

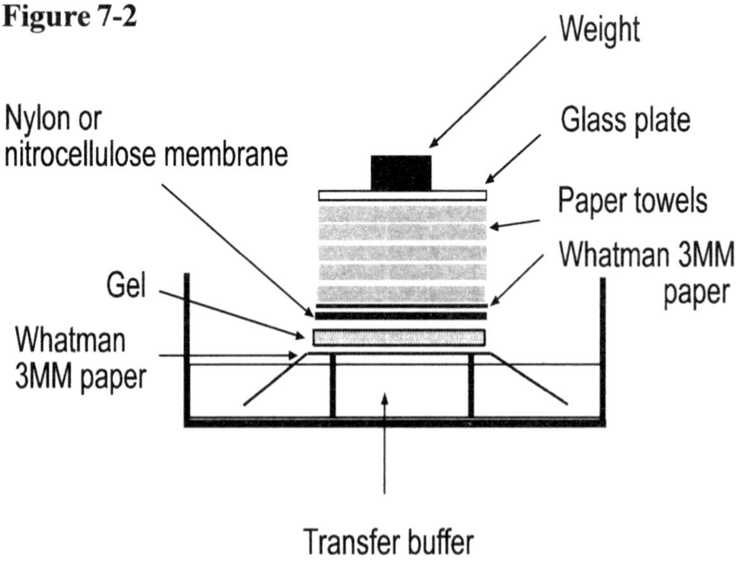

sequentially with several single locus DNA probes by a method called hybridization, where complimentary strands of DNA come together to form hydrogen bonds (Figure 7-3). One of the most common restriction enzyme used as standard for digestion is called *Hae*III.

Figure 7-3

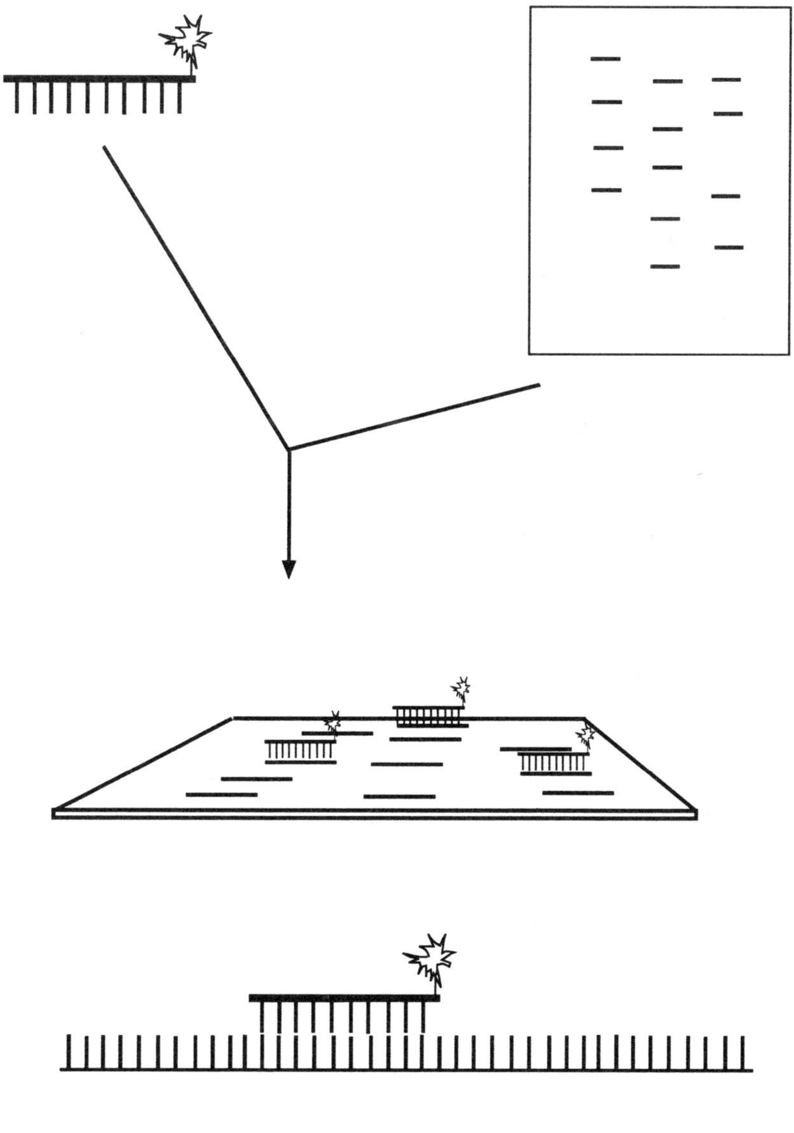

The vast majority of individuals are heterozygous at loci commonly used in DNA profiling, and thus two RFLP bands are present at one system. There are many different systems used to identify people. For example, the system that uses the *Hae*III restriction fragments, and contains a VNTR region, is called D2S44. A specific probe is used to identify the sequence and measure the allele lengths in individuals. Although the RFLP method of DNA fingerprinting is well defined and highly informative, some problems do exist. Some of these are the amount of time it takes for analysis, the effects of degraded DNA, and the amount of template DNA needed to obtain accurate and reliable results (20-100 ng). In addition, with the restriction enzyme *Hae*III, partial digestion of the template DNA may lead to extraneous bands on the gel, and the analysis format can pose problems with explanation of results. This is because extraneous bands can interfere with the analysis of multiple donors in a given relationship study.

Because RFLP suffers from various problems listed above, a new method for DNA analysis was needed. The ideal situation would be to look for smaller regions that can be used to DNA type individuals while also using very small amounts of DNA. This would be particularly useful in forensics, where samples are sometimes in very small quantities, and highly degraded, such as what you would expect from a single hair follicle or a spot of dried blood. Buccal swabs yield relatively small amounts of highly degraded and fragmented DNA. DNA that is already in poor quality may be exposed to influences like ultraviolet light, humidity, decay or storage in paraffin, under denatured conditions, leading to further degradation. Finally resulting in DNA that could never be used unless a technique were developed that could be used to examine only small parts of the genome. PCR was introduced in paternity testing practice to overcome these problems.

A survey of parentage testing laboratories in the U.S.

for 1998 showed that 36.8% laboratories use both RFLP and PCR procedures. Laboratories using PCR (STRs) only were 18.4%. RFLP is still predominating as a technique in the parentage-testing field (36.8%).

SHORT TANDEM REPEATS (STRS)

The genetic uniqueness of individuals is a central tenet of human biology. One method to examine the uniqueness of an individual is to examine short tandem repeat sequence variants (STRs) found in the genomic DNA. STRs are repeated units of sequence ranging between 2 and 7 bps. Many of these regions have been discovered to have a high degree of allele variability among the human population. The distinguishable alleles are found based on the number of repeat units (length of DNA in the region), and are amplified using PCR amplification. The human genome has an abundance of these sequence occurrences, averaging one tri- or tetra-nucleotide unit every 15 thousand bases. There have been over two thousand STR systems mapped out in the human genome, and it is estimated that thousands more remain to be discovered.

The basic function of all forensic testing, whether parentage testing or identification, is to exclude the maximun number of individuals possible. Paternity testing determines if the alleged father has the obligate allele that is identified in the case.

STRs are useful for, not only paternity testing, but also for disease diagnostics (Fragile X, Huntington's disease, Spinobulbar muscular atrophy), physical and genetic mapping, monitoring bone

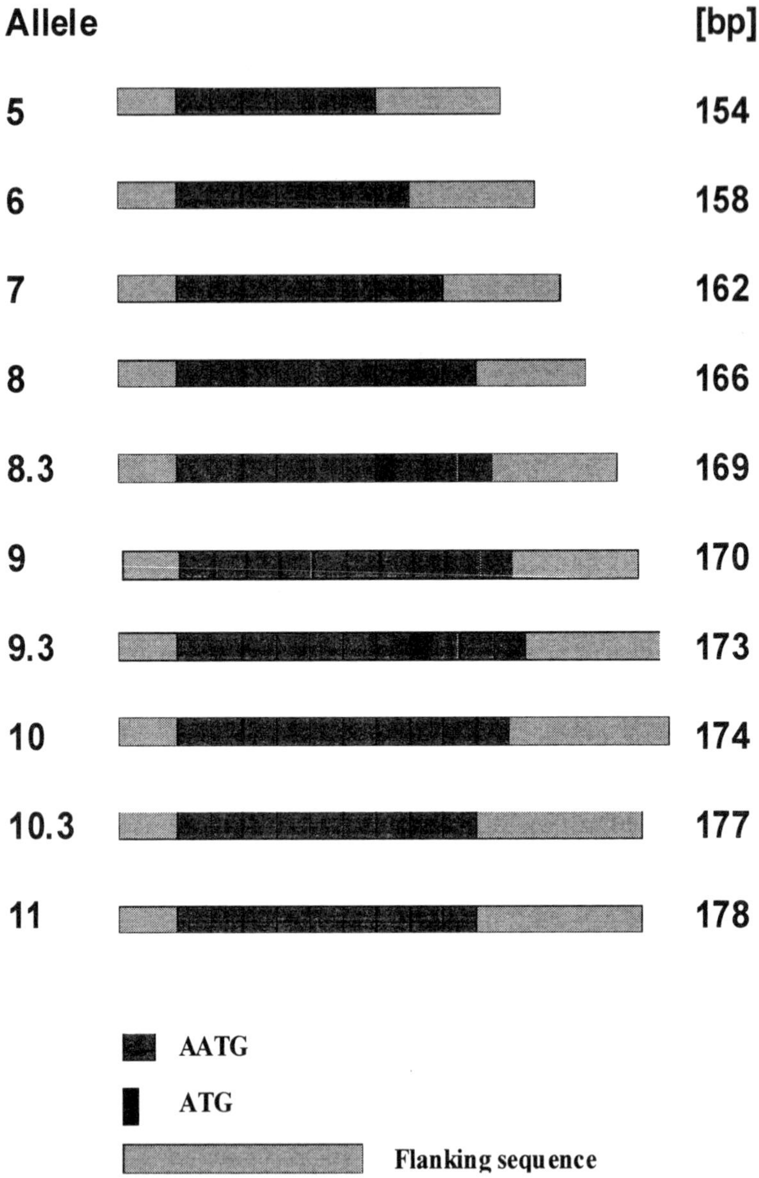
Allele
[bp]
5
154
6
158
7
162
8
166
8.3
169
9
170
9.3
173
10
174
10.3
177
11
178
AATG
ATG
Flanking sequence

Figure 7-4

marrow transplants, and forensics. Forensics and parentage assessment use STR systems because of the flow of genetic information from parent to offspring. The human genome of an offspring carries two copies of each chromosome, one copy from each parent. Chromosomes contain genes that have the DNA sequence to code for and build proteins. An exon is the region of

Table 7-1: HUMTHOl Locus

Allele Repeat #	Size [bp]	Repeat Structure
3	146	$[AATG]_3$
5	154	$[AATG]_5$
6	158	$[AATG]_6$
7	162	$[AATG]_7$
8	166	$[AATG]_8$
8.3	169	$[AATG]_5ATG[AATG]_3$
9	170	$[AATG]_9$
9.3	173	$[AATG]_6ATG[AATG]_3$
10	174	$[AATG]_{10}$
10.3	177	$[AATG]_6ATG[AATG]_4$
11	178	$[AATG]_{11}$
12	182	$[AATG]_{12}$
13.3	189	$[AATG][AACT][AATG]_8ATG[AATG]_3$

Nomenclature

Selected topics according to the DNA Commission of the International Society for Forensic Haemogenetics:

1 All loci have unique identifiers of 2-7 characters in length that are all in capital letters. For example, the Amelogenin locus is denoted AMEL.

2 Results of laboratory genetic tests are presented as the locus name followed by a space and then by the observed results (e.g., GYPA A).

3 A two gel band pattern is represented by the locus symbol, a space and symbols of the alleles seprated by a coma. For example, GYPA A,B.

4 For anonymous DNA segments such as D14S1 for example, the following rule holds:

√ the anonymous region (intron sequence) is identified as **D** for DNA segment.

√ The number after **D** represents the chromosome that the anonymous DNA is found on.

√ **S** indicates that it is a single copy region, which occurs only once in the genome. **Z** stands for multiple locations of a DNA fragment.

√ The number after **S** or **Z** indicates the order in which the DNA marker was reported as such for that chromosome.

5 Alleles or band sizes are always reported from the smallest to the largest, for example: 2.33, 3.56 kb.

6 Heterozygous phenotypes are always separated by a coma without a space.

7 Single band patterns, whether for RFLP or PCR based systems, are reported as a single band phenotype.

DNA sequence in a gene that codes for a protein. The sequence that exists between genes or a part of a gene, and does not code for a protein, is called an intron. The locations or particular positions of a gene, the intronic sequence of a gene, and the intronic sequence between genes are termed loci. A single location in the human genome is termed a locus. When a locus on a chromosome differs at the sequence level, it is called an allele. The variability contributed to STRs is caused by different numbers of repeating units or alleles (Figures 7-1 and 7-4).

One example of a STR is the locus HUMTHO1 (Figure 7-4). This particular locus is a tetrameric (4-bp repeat). Tetrameric repeat systems are used in the forensic community because PCR amplification of these systems yields fewer artifacts during DNA fragment separation, such as extra band production and repeat slippage or stuttering. The HUMTHO1 locus is located in intron 1 of the human tyrosine gene with a chromosomal position of **11**p15-15.5. The alleles present in the locus are represented in table 1, along with the bp size of each allele, and the repeat structure. The repeating unit of HUMTHO1 is (AATG). The alleles are numbered according to the number of tandem repeats present in the PCR-amplified products. If a sample has 9 **AATG** repeat motifs on one chromosome and 10 **AATG** repeat motifs on the other chromosome, then the sample is designated a HUMTHO1 (9,10). However, not all alleles in the HUMTHO1 locus carry just 4 bp repeats (AATG). Some alleles have base pair deletions (i. e., 9.3 allele in Table 7-1). The allele type is the number of 4 bp repeats. An additional 0.1 is added for each bp less than 4 sandwiched within the string of repeats. A HUMTHO1 9.3 sample has 9 of the 4 bp repeats. An extra ATG exists between the sixth and seventh repeats, accounting for the 0.3 designation in the type (0.1 for each bp of the ATG).

All STR loci are parentally inherited. One allele from each parent is passed on to the offspring. A simple example

Table 7-2 HUMTHO1 allele frequencies in United States populations

Allele	African American (N= 198)	Caucasian (N = 197)	Hispanic (N = 206)
5	0.000	0.000	0.002
6	0.111	0.221	0.228
7	0.434	0.175	0.340
8	0.189	0.127	0.083
9	0.146	0.168	0.100
9.3	0.106	0.308	0.245
10	0.013	0.003	0.002

is a father that has the type HUMTHO1 (6, 9) and a mother that has the type HUMTHO1 (7, 8). The offspring can have the following HUMTHO1 types (6, 7; 6, 8; 9, 7; or 9, 8). If the typing does not match one of these pairings, the offspring cannot be assigned to that set of parents unless the child had a mutation, which is a very rare occurance.

Another use of STRs is in forensic cases. If blood, skin, dandruff, or any other biological material is found at a crime scene, it can be analyzed using STRs and matched with the suspect's

Table 7-3 Amelogenin Locus (DNA fragment length as per Dr. Faerman's experiment.

Allele	Size [bp]
X chromosome	330
Y chromosome	218

genotype. With this use of STRs, population databases have been set up for different STR systems. The allele frequency for HUMTHO1 STR system has been analyzed for the United States populations (Table 7-2).

Heterozygosity is a way of measuring the genetic variation, which is useful in establishing the diversity that exists in a population. In the case of STR loci, the heterozygosity is the appearance of different alleles for a locus of an individual. The observed heterozygosities for the locus HUMTHO1 in different populations are 70.2% for the African American sample, 72.1% for the Caucasian sample, and 77.2% heterozygosities for the Hispanic sample.

Along with the STR systems for identification, a reliable method for sexing of samples has also been accomplished. Amplification of the X-Y homologous amelogenin gene with a single primer pair is used for a sex test because it generates different length products from the X and Y chromosomes (Table 7-3). This locus is co-amplified with STR loci making it a powerful tool in paternity testing cases. It is also used extensively in forensic science because, along with the identification power of STRs, the sex of the offender can also be known.

Marina Faerman and her colleagues at the Hadassah Medical School of Jerusalem used the gene to identify the sex of infant skeletons too young and incomplete to be classified in any other way. On an agarose gel the shorter fragment of the Y chromosome will migrate faster than the longer fragment and they are clearly distinguishable. A male skeleton will yield the two products (coming from the Y and X chromosomes) whereas a female skeleton will yield only one band (two X chromosomes).

Chapter 8

HOW TO READ A PATERNITY TEST REPORT

By H. Edward Grotjan, Ph.D.

This chapter aims to summarize all the information described in this book regarding genetic tests for paternity. Paternity tests typically analyze genomic DNA. One-half of a person's nuclear DNA is inherited from the mother and the other half is inherited from the father. Each person has two distinct copies of genetic information, called alleles, at each genetic locus (a particular location in the genome). For genomic DNA, one allele at each genetic locus is inherited from the child's mother and the other allele comes from the child's father. The general question asked in a paternity test is whether the alleles present in the child could have been inherited from the alleged father.

The human genome consists of a digital code, more than 3 billion letters in length. Each nucleated cell in the body has a full copy of that person's genetic code. The sequence of the letters in the code is particularly important because the alphabet used to encode information contains only four letters—A, C, G and T which denote the four nucleotides found in DNA: Adenine, Cytosine, Guanine and Thymine (nucleotides are a type of biochemical compound). Technically, a genetic locus corresponds to a particular segment of the genetic code that is several hundred to several thousand letters in length (also called bases or base pairs—because DNA is composed of two complementary strands). In the human genome, some genetic loci (plural of locus) have unique sequences that make them identifiable. In particular, paternity tests examine genetic loci where multiple alleles exist in the population, and the two alleles that exist in an individual are likely to be different (high degree of heterozygosity). In these circumstances, each allele typically has a unique frequency in a particular population (for example, racial group such as Caucasians, African Americans, etc).

The two general characteristics of DNA most commonly used to identify differences bewteen individuals are Restriction Fragment Length Polymorphism (RFLP) analysis and analysis of Short Tandem Repeats (STR). With RFLP analysis, alleles are usually identified by the size, in kilo bases, of the DNA fragment detected. Thus, "1.37" would denote an allele that generated a fragment 1,370 nucleotides or "letters" long when the DNA was digested with a specific restriction enzyme such as Hae III. With STR analysis, an allele is named according to the number of times a specific sequence of nucleotides (letters) is repeated. The number of repeats is determined on the basis of the size of the fragment generated by amplification of a region of the genetic locus that spans the repeated sequence. For example, the HUMTHO1 locus (human thyroid peroxidase gene) has between 5 and 11 repeats of the sequence AATG. For HUMTHO1, allele 5 has 5 AATG repeats, allele 6 has 6 AATG repeats, etc. Some genetic loci have alleles where the repeat sequence is imperfect. For example, the 9.3 allele of HUMTHO1 has 9 four-letter repeats and 1 three-letter repeat. This difference is very important because the HUMTHO1 allele designated as 9.3 occurs with a high frequency in many populations while the HUMTHO1 allele with 10 perfect repeats is fairly rare. The following discussion applies to the interpreting information obtained by either RFLP or STR analysis.

A person is termed heterozygous if they have two different alleles at a given genetic locus and called homozygous if they have two copies of the same allele. A common way to code the alleles is to denote the child's alleles with letters, that is AA if homozygous and AB if heterozygous. For example, if a child has alleles 6 and 9.3 at the HUMTHO1 locus (Table 8-1), the 6 allele would be coded as A and the 9.3 allele would be coded as B (or vice versa). Parental alleles that match the child's alleles are coded with the same letter. Thus, a HUMTHO1 6 allele in either parent would be given the code letter A while a HUMTHO1 9.3 allele in either parent would be given the code letter B (or vice versa, depending on how the child's alleles are coded). If this sounds confusing, the best way to make the conversion is to construct a table, for example:

Table 8-1 Allele Conversion Table

Allele	Code Name
HUMTHO1-Allele 6	A
HUMTHO1-Allele 9.3	B

If the child's genotype is coded as AB, the genotype of the mother must include at least one allele that matches the child. That means, her genotype must include either A or B allele. Possible matching genotypes would include AA, BB, AB, AC or

BC where C represents an allele that is not present in the child. The allele that the child inherits from its biological father is called the Obligate Paternal Allele. The alleged father must possess an Obligate Paternal Allele to be included among potential biological fathers. For example, the alleged father must possess a B allele to be included among potential fathers when a child with an AB genotype has a mother of AA or AC genotype. That is, the father's genotype must either be AB, BB, BC or BD where D represents an allele that is not present in either the child or mother. In the case where the mother and child both have an AB genotype and the father has a BC genotype, one can conclude that the A allele was maternally derived because she is the only parent that has an A allele. Sometimes there are double matches. For example if all three parties have an AB genotype, it is not possible to tell which allele is inherited from the mother and which is inherited from the alleged father. However, it is possible to state that the alleged father is included among possible biological fathers of the child.

A straightforward way to examine possible genotypes is to construct a Punnett square. In a Punnett square, parental genotypes are listed on the border of a table and the potential genotypes of their offspring are listed in a 2 row by 2 column table. Consider the case where the mother's and alleged father's genotype are AC and BD, respectively (Table 8-2).

Table 8-2

	Alleged Father (BD)	
Mother (AC)	B	D
A	AB	AD
C	CB	CD

This illustrates that it is possible for a mother who has an AC genotype to have a child with an AB genotype if the father's genotype is BD. In this example, the B allele would be the Obligate Paternal Allele and the alleged father would be excluded if he does not posses a B allele.

As an exercise, construct a Punnett square to determine if man with an AD genotype can father an AB child whose mother is AC (Table 8-3)? At first glance, it looks like both parents match the child because each has an A allele. However, under the Presumption of Maternity, the A allele is presumed to have come from the mother (rather than the alleged father)! As a Punnett square:

Table 8-3

	Alleged Father (AD)	
Mother (AC)	A	D
A	AA	AD
C	CA	CD

Thus, an AD father is excluded as the biological father when the child's genotype is AB and the mother's genotype is AC. This example illustrates that, in some respects, the genotype of the mother is given greater weight when a child's alleles are assigned to have been inherited from its mother and father.

The typical analysis for paternity involves comparing the alleles that the mother, child and alleged father have at multiple genetic loci. The reports from most paternity testing laboratories include a table listing the alleles for each party tested. If you are having difficulty interpreting a paternity testing report, constructing a Punnett square for each genetic locus is an excellent way to determine if the genotype of the alleged father matches that of the child. You should then be able to determine how the laboratory assigned the inheritance of alleles and whether an alleged father is included or excluded at a particular genetic locus. More specifically, the Punnett square analysis should tell you:

- ✓ which of the child's alleles was inherited from its mother,
- ✓ which of the child's alleles represent an Obligate Paternal Allele(s), and
- ✓ if the alleged father possesses the Obligate Paternal Allele (which translates to whether the alleged father is included or excluded).

For an alleged father to be excluded as the biological father of a child, paternity testing laboratories typically require that there be at least two genetic loci where the alleged father does not match the child. Many parentage-testing laboratories only consider an alleged father excluded when the alleged father does not match the child at three or more loci. These constraints build in a considerable safety factor to assure that the laboratory does not erroneously exclude an alleged father. However, it is very unlikely that one would find an exclusion between father and his true biological child.

A representative list of alleles observed for the mother, child and an included alleged father is given in Table 8-4. Note that maternal alleles are underlined, Obligate Paternal Alleles are

Table 8-4

Genotype of Each Person In a Paternity Test			
Genetic Locus	Mother	Child	Alleged Farther
CSF1PO	11, 10 C, A	10, 12* A, B*	12*, 11 B*, D
TPOX	9, 8 B, A	8, 8* A, A*	8*, 12 A*, C
HUM THO1	6, 9.3 B, A	9.3, 9.3* A, A*	9.3*, 6 A*, C
vWA	17, 18 C, A	18, 19* A, B*	19*, 17 B*, C
D16S539	13, 11 C, A	11, 11* A, A*	11*, 11* A*, A*
D7S820	12, 11 B, A	11, 11* A, A*	11*, 9 A*, C
D13S317	11, 11 A, A	11, 8* A, B	8*, 10 B*, C
D5S818	11, 11 A, A	11, 11* A, A*	11*, 11* A*, A*

denoted with an asterisk and letter codes (assigned as described above) are listed below each person's genotype.

In a typical analysis, a parentage testing laboratory will determine if the alleged father is excluded or included among those who could have fathered the child. Exclusions are essentially absolute. With exclusions, the chances that the alleged father is the biological father of the child are so remote that the probability is considered to be zero (0) for all practical and legal purposes. Exclusions typically are reported as a Paternity Index of 0 and a Probability of Paternity of 0%.

However, if the alleged father is included, the laboratory will perform a series of calculations to describe the level of statistical certainty associated with their analysis. To develop the concept, first consider the case where every man possesses the child's *Obligate Paternal Allele*. The Obligate Paternal Allele present in the child's genotype could have been inherited from the alleged father, but it could also have been inherited from any other man in the population. Thus, the first type of genetic match does not provide any information (i.e., is not very informative) regarding who is the biological father of this particular child. Now consider the other extreme where the Obligate Paternal Allele present in child is unique and could only have been inherited from the alleged father. The second type of genetic match is very informative and would essentially prove that the alleged father is the child's biological father. These two examples represent extremes but form the basis for calculating the statistical certainty or likelihood that an alleged father is the biological father of a particular child.

The statistical certainty that an alleged father is the biological father of a particular child is usually expressed as a Combined Paternity Index and/or Probability of Paternity. The laboratory usually calculates a Paternity Index for each genetic locus examined. The Paternity Index is a ratio of two probabilities, simplistically stated as the probability that the alleged father is the biological father of the child divided by the probability that an alternative father is the biological father of the child. Statistically, the ratio of two probabilities is called a likelihood ratio. The probability that the alleged father is the biological father is calculated using fundamental genetic principles. The probability that the child inherited the Obligate Paternal Allele from the alleged father is either 0, 0.5 or 1. Zero means the alleged father is excluded, 0.5 means that the Obligate Paternal Allele was inherited from an heterozygous father (for example, an AB child and BC father) and 1.0 means that the Obligate Paternal Allele was inherited from an homozygous father (for example an AB child and BB father). A father that is homozygous for the Obligate Paternal Allele is thus twice as strong a genetic match as a heterozygous father. The probability that an alternative father is the biological father of the child is calculated using the frequency of the Obligate Paternal Allele in the population. For most calculations, the frequency of a particular allele is taken from accumulated data corresponding to a specific racial group (the reason that race will usually be noted at the time of specimen collection). The rarer the Obligate Paternal Allele in the population, the more likely that the child inherited the allele from an included alleged father and the higher the Paternity Index. For example, consider the CSF1PO locus in Table 8-5. The probability that the child inherited allele 12 from the alleged father is 0.5. Allele 12 is found with a frequency of 0.330 (33.0%) in the racial group (Caucasians) of the alleged father. Hence, the Paternity Index is 0.5 divided by 0.33 or 1.515. The *Combined Paternity Index* is obtained by multiplying the Paternity Indices from each genetic loci under the assumption that each (statistically) represents an independent event. In the table below, the *Combined Paternity Index* of 15, 324 means that theoretically you have to test 15,324 unrelated men chosen at random from the same racial population to find a second person that would be a perfect genetic match with the child. The higher the *Combined Paternity Index*, the more likely that the alleged father is the biological father of the child. The *Combined Paternity Index* is a statement regarding the strength of the genetic evidence.

Table 8-5

Genetic Locus	Mother	Child	Alleged Father	Paternity Index
CSF1PO	11, 10	10, 12*	12*, 11	1.515
TPOX	9, 8	8, 8*	8*, 12	0.947
THO1	6, 9.3	9.3, 9.3*	9.3*, 6	1.511
vWA	17, 18	18, 19*	19*, 17	5.747
D16S539	13, 11	11, 11*	11*, 11*	1.567
D7S820	12, 11	11, 11*	11*, 9	1.695
D13S317	11, 11	11, 8*	8*, 10	3.497
D5S818	11, 11	11, 11*	11*, 11*	2.710
F13A01	4, 7	7, 6*	6*, 4	1.742
F13B	13, 10	10, 6*	6*, 9	5.000
FESFPS	13, 10	10, 11*	11*, 12	1.139
LPL	10, 11	11, 12*	12*, 12*	4.926
Combined Paternity Index				15,324.5
Prior Probability (%)				50.0
Probability of Paternity (%)				99.9935

Furthermore, the Probability of Paternity is calculated on the basis of genetic and non-genetic evidence. The non-genetic evidence is included in the calculations as the Prior Probability. Most laboratories use a neutral Prior Probability of 50% when calculating the Paternity of Paternity but can be instructed by a court of law to use any value between 0 and 1 (0 and 100%). If the Prior Probability is set to zero, the alleged father is excluded irrespective of the genetic evidence and if set to 1.0, the alleged father is legally proven to the biological father irrespective of the genetic evidence. Note that the Probability of Paternity never reaches a value of 1 (or 100%) when the Prior Probability is less than one. But, when a large number of genetic loci are examined, the Probability of Paternity comes very close to 1 in the form of a long string of 9's after the decimal point (99.99999% or higher). From a practical perspective, most laboratories quit examining additional loci when they reach some pre-determined Combined Paternity Index or Probability of Paternity. If additional loci were examined, the string of 9's would simply get longer. Sometimes this is important. For example, we performed a test that involved three alleged fathers that were themselves related as a father and his two sons. Thus, it was important that we unequivocally identify the biological father of the child by examining a sufficient number of loci to exclude the two non-fathers. Some laboratories may provide other statements of statistical certainty, such as a "random man not excluded" to provide further detail in reporting inclusions.

Most states set a minimum legal standard, typically requiring a Probability of Paternity of 99% that must be exceeded to legally establish that a particular person is the biological father of a child. The 99% value should be viewed as a minimum standard. It means that 1 person in 100 chosen at random from the same racial population would match the child's genetic profile. Most

EXAMPLE

A Paternity Analysis Report contains the following table (Table 8-6*). Please comment on the results.

Table 8-6

Case 78543 Name DoB - Race Date Collected Test No.	**MOTHER** Susan Smith 6/20/64 Caucasian 2-12-00 78543-10	**CHILD** Peter Smith 10/13/89 Caucasian 2-12-00 78543-20	Alleged **FATHER** Derek Malloy 6/26/62 Caucasian 2-12-00 78543-30
Results	**Allele Sizes**	**Allele Sizes**	**Allele Sizes**
D2S44 YNH24 - *Hae III* PI 19.20	2.53 1.56	2.53 1.26	1.89 1.26
D4S139 pH30 - *Hae III* PI 32.70	5.03 4.56	5.03 5.48	5.49 3.47
D10S28 TBQ7- *Hae III* PI 9.70	5.44 3.35	5.44 4.20	5.21 4.20
D17S26 EFD52 - *Hae III* PI 73.80	4.43 4.20	4.43 2.68	4.29 2.68

Comments: The laboratory used four probes to carry out the analysis, i.e. YNH24, pH30, TBQ7 and EFD52. These probes were labeled and applied to the membrane carrying genomic DNA that was digested with the enzyme *HaeIII*. The probes were for the four loci D2S44, D4S139, D10S28, and D17S26. The first locus is an intronic sequence (**D** for automous DNA) on chromosome 2. This sequence is a single copy region (**S**) and was the 44th marker discovered. After the analysis there were two bands on the gel - 2.53 and 1.56 kb for the mother. The father produced 1.89 and 1.26 kb bands. The child had 1.26 kb band coming from the father (the mother does not have this band) and one 2.53 kb band corresponding to the mother's profile. The same pattern of inheritance was observed for the other three markers.

*The following example is a courtesy of *DNA Diagnostics Center* (Fairfield, OH).

Therefore the conlcusion is that the alleged father is very likely to be the biological father of the child. Additionally the reported PIs give 19.20 x 32.70 x 9.70 x 73.80 = 449445.54. This means that the alleged father is 449,446 times more likely than a random man of the same race to be the father of the child.

EXERCISES FOR THE READER

I. A Paternity Analysis Report contains the following table (Table 8-7). Please comment on the results.

Table 8-7

Case 78547 Name DoB - Race Date Collected Test No.	**MOTHER** Dianne Bowers 2/10/64 Caucasian 5-12-00 78547-10	**CHILD** Joseph Bowers 08/11/89 Caucasian 5-12-00 78547-20	Alleged **FATHER** Kirk Schulz 6/07/62 Caucasian 5-12-00 78547-30
Results	**Allele Sizes**	**Allele Sizes**	**Allele Sizes**
D2S44 YNH24 - *Hae III* PI 0.0	2.53 1.56	2.53 1.26	3.97 2.23
D4S139 pH30 - *Hae III* PI 0.0	5.03 4.56	5.03 5.48	2.21 1.84
D10S28 TBQ7- *Hae III* PI 0.0	5.44 3.35	5.44 4.20	6.81 2.25
D17S26 EFD52 - *Hae III* PI 0.0	4.43 4.20	4.43 2.68	3.62 3.16

Reader's Comments and Notes:

II. Some of the data from a Paternity Analysis Report came from this autoradiogram of a gel (Fig. 8-1). Please comment on the results.

Figure 8-1

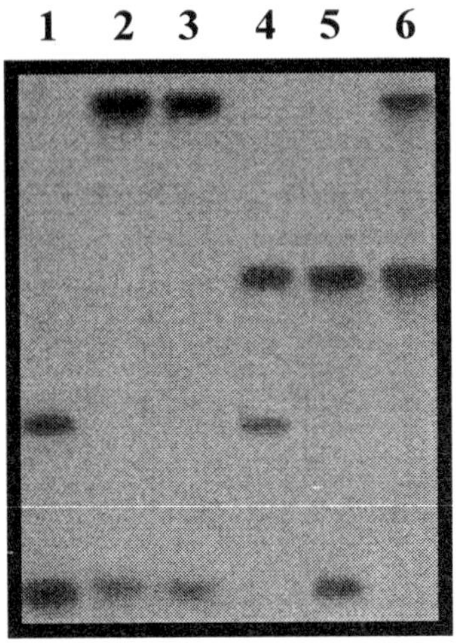

Lane	Family Member
1	Child 1
2	Child 2
3	Mother
4	Father
5	Child 3
6	Child 4

Reader's Comments and Notes:

Chapter 9

SIBLINGSHIP AND GRAND-PARENTAGE STUDIES

DNA analysis can also provide information about relationships such as siblingship or grand-parentage. DNA of two individuals can provide even more information, such as whether they are full or half siblings. However, the answers provided by siblingship testing are nearly as conclusive as paternity testing can be.

Because the parents have two copies of each of the 23 chromosomes, there is a 50% chance that they will randomly pass on a particular chromosome to their offspring. If two siblings share the same mother and father, theoretically, they should share 50% of their mother's chromosomes, and 50% of their father's chromosomes. Therefore, during a siblingship test, many different chromosomes are analyzed. If two people are full siblings, 50% of their alleles should be identical. If two people are half siblings 25% of their alleles should be identical. By comparing several different alleles, the numbers of shared alleles are analyzed, and a siblingship index is calculated. The siblingship index indicates the probability that a random person in the population would have the same-shared alleles as what were established.

Since there are no alleles that must match between full siblings (with the exception of the Y-chromosome and X-chromosome), there is the possibility that none of the alleles examined are actually shared by two siblings. Therefore, it cannot be concluded with 100% accuracy that the two alleged siblings are or are not true siblings.

If the siblings are known to have the same mother, DNA testing can be used to increase the power of discrimination for a siblingship analysis. Additionally, by testing other paternal relatives (i.e., grandparents, aunts and uncles) the power of discrimination also can be increased.

MORE DISTANT RELATIONSHIP STUDIES INVOLVING THE Y-CHROMOSOME AND MITOCHONDRIAL DNA

Of the 23 pairs of chromosomes that make up the human genome, a female has a pair of X-chromosomes (female gender) and a male would have one X and one Y chromosome (male gender). The Y-chromosome is passed on from father to son (Figure 9-1).

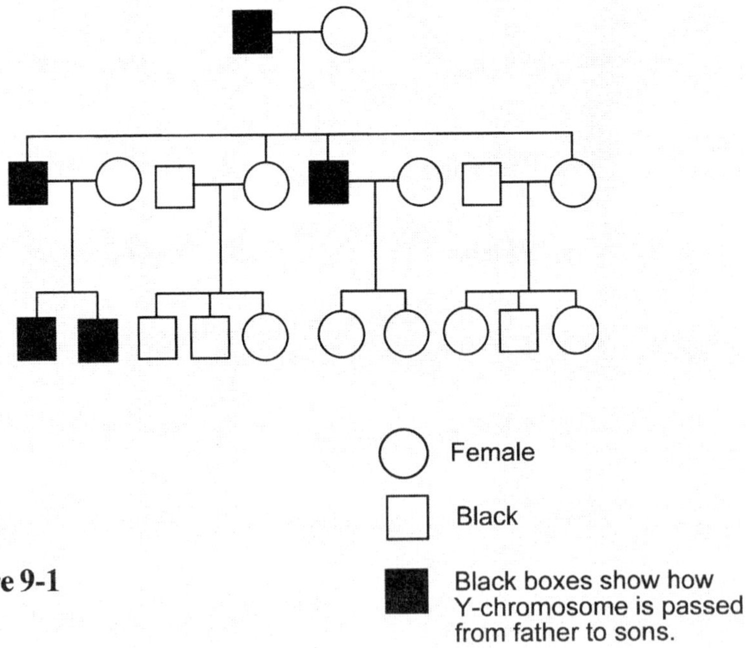

Figure 9-1

In addition to chromosomes, there is extra-chromosomal DNA found in the cell called mitochondrial DNA (mtDNA). This DNA is always passed from the mother to all the children (Figure 9-2). Y-chromosome and mtDNA can be used as a useful tool for studying more distant family relationships.

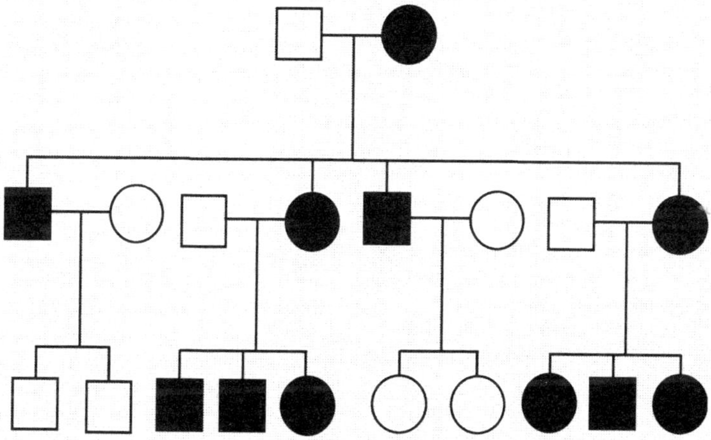

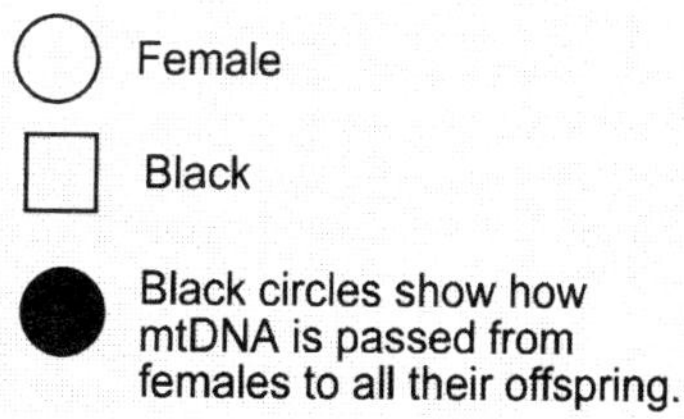

Figure 9-2

THE Y-CHROMOSOME TO TRACE OUR FAMILY NAMES, ETHNICITY, RELIGION, AND ORIGIN

According to Chapter 5 of *Genesis* Adam begat Seth, and Seth begat Enosh, Enosh begat Kenan, and so on.... Translated in Y-chromosome genetics terms, Adam passed a copy of his Y chromosome to Seth, and Seth passed a copy of his Y-chromosome to Enosh, and Kenan inherited the

The Y chromosome can provide important information about lineages from a specific male. It is present in normal males as a single copy that is paternally inherited and can not recombine or rearrange because there is only one Y chromosome.

chromosome from Enosh. According to this Biblical account, the Jewish priesthood was established 3,300 years ago and the first Israeli high priest was documented. Designation of Jewish males to the priesthood continues to this day. This spiritual lineage is determined by strict patrilineal decent. This lead Michael Hammer of the University of Arizona in Tucson and his collaborators to construct patrilineal genealogy cladrograms by using the Y-chromosome feature of being passed from father to son. Molecular biology techniques are being currently refined to unravel the mysteries of patrilineal inheritance.

The Y-chromosome has another important characteristic, i.e. it does not recombine with the X chromosome (with the exception for the pseudo-autosomal region) and it seems to remain quite conserved in its other regions over many generations. It is noteworthy that the human X and Y chromosomes, like those of other animals, have evolved from an ordinary pair of autosomes. It has been postulated that the beginnings of the X-Y differentiation occurred about 240 to 320 million years ago, shortly after the divergence of the mammalian and avian lineages. The non-recombining regions of the X and Y chromosome have become highly differentiated during evolution.

The Y-chromosome has a low mutation rate. Therefore, mutations on the Y-chromosome represent a record of its evolutionary past that can be used in genealogical tests to map out the diversity of populations and/or groups. As long as a mutation does not affect the individual's ability to reproduce, it may be preserved and handed down to offspring. An exchange of a single DNA building block, i.e. a nucleotide, with another is called a Single Nucleotide Polymorphism (SNP). Different combinations of polymorphisms are known as haplotypes. By looking at these changes, biological relationships between two or more individuals can be established. SNPs are an alternative to STRs for examining Y-chromosome inheritance.

Peter Underhill of Stanford University has used Y choromosome SNP markers to discover genetic evidence that modern humans are descendants from a single migration of archaic Homo sapiens out of Africa. Underhill considers the Y chromosme as a "playground to study human history".

The Y chromosome has also been used to study the colonization of Europe and the origin of some languages by employing a method used to analyze microsatellite markers (STRs). The Finnish language is unique in Northern Europe in tracing its ancestry to the Uralic, rather than the Indo-European language family. This suggests that some of the Finnish genome originated in Northern Asia. But earlier genetic analysis showed that Finnish populations were closely related to the rest of Europe. Scientists have supported the hypothesis that the spread of language, not genes, accounts best for the uniqueness of the Finnish. However, Chris Tyler-Smith's group at the University of Oxford has discovered that Y polymorphisms present in abundance in Asia are also widespread in Finnish populations. They presented evidence of a migration out of Asia that is hard to refute. Half of the Uralic-speaking Finns' Y chromosome is related genetically to central Asian Uralic speakers. The study conforms to the theory that the passing of a language from generation to generation is governed more by patrilineage (by father) than by matrilineage (by mother).

Another correlation study between the Y chromosome and linguistic heritage was carried out in Basques, and geographically surrounding populations. The Basques have a unique language, which is very different from the languages of surrounding ethnic groups. Tyler-Smith's group and collaborators used Y-chromosome markers to study if Basques' Y-chromosomes are distinct from those of surrounding populations. They demonstrated a specific Y-chromosome lineage, which has a recent origin and is rare or absent in most other parts of the world. However, this Y-chromosome lineage is shared with high frequency between Basques and Catalans, who speak languages belonging to different language families. These researchers presented evidence and concluded that there has been male-mediated gene flow directly between Basques and Catalans since the establishment of the languages spoken by them now. The polymorphism they studied occurred a few thousand years before the present. Interestingly, the Basques Y-chromosome lineages were also found in South America, which is not unexpected due to recent historical analysis about the ancient links between Iberia and South America. Markers were also found in France, Germany and England.

Y-chromosome fingerprinting can be useful to genealogists to study surnames and the history of their ancestry. Surnames in most European populations came into existence after the 13th century. Since the English Mediaeval period it has been common practice for children to take their father's surname. Bryan Sykes of the University of Oxford decided to study the correspondence between the surname "Sykes" and

Y-chromosome haplotype. The highest residential concentrations of Sykes are in the counties of West Yorkshire, Lancashire, and Cheshire. Bryan Sykes and his colleague reported a highly significant association between Y-chromosome haplotype and the Sykes sample. This work has very important forensic and genealogical implications because such a method will be used to investigate the link between branches with the same surname for which a common ancestor could not be established from written records.

The Y-chromosome became popular in response to a DNA paternity testing case that involved US President Thomas Jefferson. The publication on this research was released in the scientific magazine *Nature* by Eugene Foster, a retired pathologist in Charoltsville, Viriginia and co-authors from England.

In 1802, President Thomas Jefferson was accused of having fathered a child by one of his slaves, Tom Woodson, whose mother was Sally Hemings, was born in 1790, after Jefferson and Sally Hemings returned from France. Members of the African-American Woodson family believe that Thomas Jefferson was the father of Thomas Woodson, whose name comes from his later owner. The scientific team had the challenging objective of determining if descendants from Thomas Woodson were related to President Thomas Jefferson.

Sally Hemings had at least four more children. Her last son, Eston was born in 1808. Records point out that he had striking resemblance to Thomas Jefferson, which gave him access to white society in Madison, Wisconsin as Easton Hemings Jefferson. Eston's descendents believe that Thomas Jefferson was his father. Scholars who have devoted their efforts to studying President Jefferson's life maintain that Samuel or Peter Carr, sons of Jefferson's sister, fathered Sally Heming's latest children including Eston.

The Foster team analyzed samples from five male-line descendants of two sons of Thomas Woodson, one male line-descendent of Easton Hemings Jefferson, and three male-line descendents of three sons of John Carr, grandfather of Samuel and Peter Carr. They reported the most probable explanation of their findings was that Thomas Jefferson, rather than one of the Carr brothers, was the father of Eston Hemings Jefferson. The researchers concluded that Thomas Woodson was not Thomas Jefferson's son.

The published data were both scientifically challenged and took on a political spin. The challenge came from Herbet Barger of Fort Washington, Maryland, a genealogist and husband of a Jefferson family descendent. Mr. Barger helped locate living members of the Jefferson family and persuaded them to donate blood for the DNA study. However, he was not acknowledged in the article and his theory was not mentioned. Barger argues that the most likely farther of Eston Hemings is not Thomas Jefferson, who was 65 at the time Eston was conceived, but Jefferon's brother Randolph. The latter lived 20 miles away and was 12 years younger than Thomas Jefferson. Barger also theorizes that Randolph's sons are candidates too. One of them Isham was reported to having parties in the slave quarters at night.

The Y-chromosome is important not only for genealogical purposes. It contains information for the expression of many traits. However, the lack of knowledge to correlate the differences among Y-chromosome and phenotypic differences among males makes this field *terra incognita* (unknown land) until now. A trait, which is controlled by a locus found only on the Y-chromosome is termed holandric. With the exception of azoospermia and possibly gonadoblastoma, pure holandric inheritance in humans is unknown. David Goldman of the National Institute of Health (NIH) and collaborators from Finland were able to present evidence that differences among Y-chromosomes contribute to variation in vulnerability to alcohol depenedence. However, the authors point out, these differences do not demonstrate an association between Y haplotype and the personality variables thought to underlie the subtypes of alcoholism.

Several companies use Y-chromosome protocols to

offer genealogical tests. Samples are collected by buccal swabs and the procedures are as described in previous chapters. The challenging part of such services stems from the very close interaction between the users of the test and the researchers. As much information as possible is needed to couple DNA results with historical data.

MITOCHONDRIAL INHERITANCE AND DISTANT FAMILY RELATIONSHIP APPLICATIONS*

Traditional inheritance views the nucleus as the central repository of genetic information and meiosis as the principal determinant of the segregation of traits in families. However, the existence of another genome, the mitochondrial genome, in all cells introduces a twist of biology leading to nontraditional inheritance.

Mitochondria are organelles that provide much of the energy cells use for the work they do. Most biologists now believe that these structures evolved from microorganisms that established symbiotic relationships with the ancestors of animal cells very early in the history of life on this planet. Selection for metabolic advantages gained through symbiosis explains how it has come to be that mitochondria contain their own DNA that codes for 13 of their proteins along with ribosomal and transfer RNA that

MtDNA is maternally inherited, and represents the the female ancestry of an individual.

specifically help express mitochondrial series. Mitochondrial DNA (mtDNA) has also a non-coding region called the displacement loop (D-loop) or control region, which is approximately 1100 bp in length.

*This chapter contains excerpts from Butler and Levin paper, 1998. The original paper is a contribution of US National Institute of Standards and Technology and is not a subject to copyright.

Figure 9-3

The use of mtDNA sequence information to identify the remains of the Russian Tsar Romanov's family (Fig. 9-3) has received considerable public attention. Although their bones were in the ground for over 70 years, an exact mtDNA sequence match of the D-loop region was made between the Tsarina, her three daughters and the (living at the time of the study) Prince Philip of the UK whose maternal grandmother was the Tsarina's sister. Tsar Nicholas II was identified by comparison of the mtDNA sequence from the bones in the grave with that of the great-great-great granddaughter and great-great grandson of the grandmother of Tsar Nicholas II, Louise of Hesse Cassel. The match was exact except for one site at which the putative Tsar's mtDNA had both C and T bases. How is it possible to have both bases? This is explained by the phenomenon called heteroplasmy.

As with any form of DNA, mtDNA sequences are susceptible to mutation. In fact, there is evidence that mitochondrial sequences may mutate at rates 3 to 5 times greater than nuclear sequences. The consequences of mitochondrial mutations, however, may be very different from those that occur in nuclear DNA. First, each cell contains about 100,000 mitochondria, each of which has 2 to 10 copies of its genome. The effect a mutation in mtDNA will have on a cell's function will therefore depend on the number of mutant organelles in a cell compared to the number of normal, or "wild type", present. In this respect, each cell is analogous to an organism in which somatic mutation can produce mosaicism. Here the mixture of genotypes is termed heteroplasmy.

When cells divide, their mitochondria independently replicate and then distribute randomly into daughter cells. This leads to variable phenotypes within and among tissues ranging from non-viable cells (hence death of tissue), to energy generation dysfunction, to sub-threshold changes (i.e. "silent" mutations) that do not affect overall cell function. The latter event describes the

MtDNA plays an important role in identification of human remains, particularly skeletal, or decomposed bodies.

condition of the royal mitochondrial profile.

To confirm the match of the Tsar Nicholas' mtDNA with that of Louise of Hesse Cassel, the researchers took samples from Georgij Romanov, the Tsar's brother, and sequenced his mtDNA. Notably, the scientific team found the same heteroplasmy. This information and the parentage information obtained from STR analysis enabled the forensic investigators to be 98.5% certain that the bones were those of the Romanovs.

In a similar fashion, the mtDNA information from Anna Anderson Manahan, who had claimed to be the missing Royal Duchess Anastasia, showed that the claim was false. Each year parentage testing laboratories will receive a few calls from people who would like to test their DNA relationship to the Royal family. In most of the time it is not a straightforward job and a lot of efforts are usually involved.

Today, mtDNA typing is utilized in cases of distant family relationships and in which the nuclear DNA is too degraded or can not be recovered in sufficient quantities to be typed. Several laboratories in the US are involved in using mtDNA for forensic identifications. Human skeletal remains from the Vietnam War and other wars are routinely being analyzed and identified by mtDNA typing. The success of the Armed Forces DNA Identification Laboratory in this endeavor has led to the formation of a repository of blood samples from all those entering the US Armed Forces, so that, if necessary in the future, the DNA in the stored blood samples can be used to identify war casualties. Currently, because no repository existed for past soldiers, family reference samples from siblings or maternal relatives are used to confirm the identity of the human remains. Skeletons recovered from the ten-year-old Guatemalan mass graves have also been successfully analyzed by mtDNA testing. The same approach has been applied for graves in Kosovo and other mass murders worldwide.

IMPORTANCE OF MtDNA TYPING

Mitochondria are passed from generation to generation only through maternal egg cells where they are abundant. Those present in sperm are concentrated in the tail and do not contribute to the compliment of the fertilized zygote. Molecular genetic tracking of polymorphic or variable regions through families and even across the millennia of human evolution confirm maternal inheritance of mitochondrial DNA

A number of specific disorders have been described where mitochondrial mutations are either inherited or occur early enough in development to dominate most cells. These disorders characteristically affect muscle and nervous tissue, particularly the optic tracks. Among these conditions are Leber's Hereditary Optic Neuropath (LHON), which usually presents with onset of symptoms after puberty; Kearns-Sayre syndrome and chronic progressive external ophthalmoplegia (CPEO), which both result in paralysis of external, but not internal, eye muscles; myoclonic epilepsy with ragged red fibers (MERRF), which presents at various ages; and mitochondrial myopathy, encephalopathy, lactic acidosis with stroke-like episodes (MELAS), which evolves during infancy.

Although mitochondrial inheritance looks easy to identify in a pedigree, it must be remembered that phenotypic variability is a hallmark for these disorders, especially in cases where there is heteroplasmy. Therefore, a family history where some individuals present with strokes, others with muscle weakness, and still others with psychiatric or ophthalmologic complaints might not be immediately recognizable as one expressing a single mitochondrial mutation. This is an important point because while the human mitochondrial genome has been sequenced, a complete spectrum of its viable mutations is far from assembled.

Maternal inheritance looks very much like mendelian autosomal dominant inheritance with two important exceptions. First, all maternal offspring are usually affected. Even in highly penetrant autosomal dominant disorders only 50% of offspring are expected to be affected. Secondly, mitochondrially-inherited traits are never passed through a male. Therefore, males are as likely to be affected as females, but their offspring are not at risk.

With frequencies approaching 1:50,000 LHON and other mitochondrial disorders would appear to be rare. It is likely, however, that since mitochondrial mutations are frequent, there are many new phenotypes yet to be discovered. In aggregate, clinically significant mutations may be eventually shown to be common. Indeed, it has been proposed that accumulating mitochondrial mutations contribute significantly to aging.

A final point about mitochondrial inheritance that needs to be stressed is that most of the proteins contributing to mitochondrial structure and function are encoded by nuclear genes and their mutations therefore segregate as Mendelian traits. Thus, a clear distinction needs to be kept between mitochondrial disorders, most of which are inherited as recessive traits, and mitochondrial inheritance, which is concerned only with those mutations occurring in mitochondrial DNA. Disorders of mitochondrial function inherited by either classical Mendelian or mitochondrial inheritance may have overlapping features.

METHODOLOGY OF MtDNA TYPING

Forensic identifications using mtDNA start by using PCR to amplify several-hundred base pairs in each of the hypervariable regions. A variety of PCR primers have been used in the process. Fluorescent automated sequencing is typically used to examine PCR-amplified material, and the sequencing information is confirmed by analysis of both forward and reverse DNA strands or by multiple sequencing of the same strand. The US National Institute of Standard and Technology has developed a human mtDNA Standard Reference Material that can be used for quality control when sequencing all or part of the mtDNA.

Depending on the primers used, sequencing the HV1 and HV2 regions provides sequence information for positions 16,000 - 16,430 and 40 - 440, respectively. The results are then usually reported in terms of their variation from the Anderson reference sequence. The control region of the mtDNA has been estimated to vary about 1-35 (1-3 nucleotides out of every 100 is different) between unrelated individuals, and comparison of the sequences of the control region from many individuals has indicated that difference from the Anderson sequence may occur at most of the positions across the region.

METHODS FOR MATCHING SAMPLES

The uniqueness of a specific test sample may be determined by comparing the sequence information from the HV1 and HV2 regions to that provided in population-specific databases. Some of this information is available through the World Wide Web (http:/www.gen.emory.edu/mitomap.html). The significance of a match between samples is commonly determined by the counting method, in which the frequency of mtDNA type is defined as the number of occurrences of the sequence in the whole database.

The expense and time necessary to carry out PCR amplification and fluorescent automated sequencing may prohibit many forensic laboratories from becoming involved in mtDNA typing. However, with the progress of the sequencing technology it might not be such a bottleneck.

Of primary concern for the genetic testing or forensic laboratory is the possibility of contamination, especially when samples are available in limited amounts , are old or are partially degraded. Extreme care is taken when extracting the DNA in mtDNA-testing laboratories to avoid the problem of contamination. Another technical difficulty often encountered is the presence in some individuals of a C-stretch region, in which a T to C transition in position 16,189 can result in a stretch of 10 to 13 or more cytosines on the light DNA strand. This uninterrupted C-stretch region presents technical difficulties for accurate sequencing results.

EXAMPLES

Several examples of reports on Y-chromosome (Figure 9-4), mtDNA (Figure 9-5), and Full-siblingship (Figure 9-6) analyses are shown on the next following pages.

GeneTree Inc.
3150 Almaden Expressway, #203
San Jose, CA 95118
http://www.genetree.com/

Toll free: (888) 404-GENE
Phone: (408) 723-2670
Fax: (408) 723-2671

Y-chromosome Analysis Report

(GTYU: ID 00001-001)

Sample Tested (04/25/00):

Specimen Sample: John Doe
Specimen Sample: John James Doe

Services Requested: ☒DNA Profile/Relatedness ☐Forensic ☐Clinical ☐Diagnostic
Case type: ☒Personal ☐Civil ☐Criminal ☐Other:________________

Explanation of DNA Profile:
Four discrete locations on the Y-chromosome were used to construct the DNA profile for John Doe and John James Doe. Table 1 shows John Doe to have identification marker 2 at the y388 Y-chromosome location. John James Doe has the identification marker 3 at this location. This suggests that John Doe and John James Doe are excluded as having the same paternal inheritance. Results from other identification markers are also listed in Table 1. Each marker has an established probability of occurrence in the general population and among different ethnic groups.

DNA Profiles:

Table 1: The DNA profiles of John Doe and John James Doe.

Y-loci:	y388	y390	y393	y156
John Doe	2	4	2	2
John James Doe	3	3	1	3

Conclusion:
DNA profiling was performed by the PCR method, and has been completed on samples in the name of John Doe and John James Doe. Based on the frequency of the alleles determined, and matching between John Doe and John James Doe, we calculate that there is 0% probability that John Doe and John James Doe are from the same paternal lineage.

Sample Type: ☐Blood ☐Bone ☒Saliva-Buccal Swab ☐Buccal Brush ☐Hair ☐Semen ☐Tissue ☐Tooth pulps

____
Senior Clinical Scientist

Figure 9-4

GeneTree Inc.
3150 Almaden Expressway, #203
San Jose, CA 95118
http://www.genetree.com/

Toll free: (888) 404-GENE
Phone: (408) 723-2670
Fax: (408) 723-2671

Maternity Analysis Report

(DGT-00001X)

Samples Tested (04/25/00):

Specimen Sample #1: John Doe Jr.- Child
Specimen Sample #2: Jane Doe- Alleged Mother

Services Requested: ☒DNA Profile/Relatedness ☐Forensic ☐Clinical ☐Diagnostic
Case type: ☒Personal ☐Civil ☐Criminal ☐Other:________________

Explanation of DNA Profile:
Four discrete locations on several different chromosomes were used to construct the DNA profiles for each individual tested. Table 1 shows John Doe Jr.- Child to have identification markers 8.24 and 4.32 at the D4S163 STR chromosome location, and Jane Doe- Alleged Mother to have identification markers 7.54 and 4.32 at the same STR chromosome location. This suggests that identification marker 4.32 was passed from the alleged mother to the child. Results from other identification markers are also listed in Table 1. Each marker has an established probability of occurrence in the general population and among different ethnic groups.

DNA Profiles:

Table 1: The DNA profiles of John Doe Jr.- Child and Jane Doe- Alleged Mother.

Samples:	D2S44	D4S163	D7S467	D10S28
John Doe Jr.	3.32	8.24,4.32	5.36,3.85	2.00,1.51
Jane Doe	3.32,1.57	7.54,4.32	4.45,3.85	4.70,2.00

Probability of maternity: >99%

Conclusion:
DNA profiling was performed by standard methods, and has been completed on samples in the name of John Doe Jr.- Child and Jane Doe- Alleged Mother. Based on the observed scientific evidence, it is concluded, for all practical purposes, in reference to the submitted samples, that there is >99% probability that Jane Doe is the biological mother of John Doe Jr.-Child. Therefore, Jane Doe- Alleged Mother is **included** as the possible biological mother of John Doe Jr.-Child.

Sample Type: ☐Blood ☐Bone ☒Saliva-Buccal Swab ☐Buccal Brush ☐Hair ☐Semen ☐Tissue ☐Tooth pulps

Senior Clinical Scientist

Figure 9-5

GeneTree Inc.
3150 Almaden Expressway, #203
San Jose, CA 95118
http://www.genetree.com/

Toll free: (888) 404-GENE
Phone: (408) 723-2670
Fax: (408) 723-2671

Full-Siblingship Analysis Report
(DGTL-00001X)

Samples Tested (04/25/00):
Specimen Sample #1: John Doe- Sibling A
Specimen Sample #2: James John Doe - Sibling B

Services Requested: ☒DNA Profile/Relatedness ☐Forensic ☐Clinical ☐Diagnostic
Case type: ☒Personal ☐Civil ☐Criminal ☐Other:________________

Explanation of DNA Profile:
Six discrete locations on several different chromosomes were used to construct the DNA profiles for each individual tested. Table 1 shows John Doe- Sibling A to have identification markers 4.55 and 3.45 at the D6S132 STR chromosome location, and James John Doe - Sibling B to have identification markers 4.55 and 3.45 at the same STR chromosome location. Each identification marker is handed down from parents to children. One parent will provide one of the identification markers and the other parent will provide the other identification marker. Each marker has an established probability of occurrence in the general population and among different ethnic groups.

DNA Profiles:

Table 1: The DNA profiles of John Doe- Sibling A and James John Doe - Sibling B.

Samples:	D2S44	D4S139	D4S163	D5S110	D6S132	D7S467
John Doe	1.73	12.07, 5.62	5.56, 4.54	3.81, 1.42	4.55, 3.45	4.47, 3.85
James John Doe	2.01, 1.73	6.47, 5.62	5.56, 2.07	3.81, 2.18	4.55, 3.45	4.47, 3.85

Probability of full-Siblingship: >99.9998%

Conclusion:
DNA profiling was performed by the PCR method, and has been completed on samples in the name of John Doe- Sibling A and James John Doe- Sibling B. Based on the observed scientific evidence, it is concluded, for all practical purposes, in reference to the submitted samples, that there is >99.9998% probability that James John Doe- Sibling B and John Doe- Sibling A are full-siblings. Therefore, John Doe- Sibling A is **included** as the biological sibling of James John Doe- Sibling B. The likelihood that they share the same biological mother and father is 626,020 to 1.

Sample Type: ☐Blood ☐Bone ☒Saliva-Buccal Swab ☐Buccal Brush ☐Hair ☐Semen ☐Tissue ☐Tooth pulps

Senior Clinical Scientist

Figure 9-6

EXERCISES FOR THE READER

I. A Y-chromosome Analysis Report contains the following table (Table 9-1). Please comment on the results.

Table 9-1

Y-loci	Sam Nilkuk	Ivan Nilkuk
DYS288	116	116
DYS289	148	148
DYS310	173	177
DYS384	354	357
DYS385	360	360
DYS389	250	250
DYS390	214	214
DYS391	281	281
DYS392	361	361
DYS393	120	120
DYS395	118	118
GGAAT1B07	186	186

Reader's Comments and Notes:

Chapter 10

SEROLOGICAL METHODS USED TO DETERMINE BIOLOGICAL RELATIONSHIPS

BLOOD TYPING (*ABO*)

The blood typing technique for determining biological relationships is based on a cellular approach and not on DNA technology. It has been discovered that each biological species has some cell surface proteins (glycoproteins) called antigens. These molecules are typical for a species and foreign to other species. More importantly, within a given species, individuals may also differ from one another in the antigens they possess.

The technique is rather simple. The anigens present on a person's red blood cells determine his or her blood types. Every antigen has a corresponding antibody and this reaction of antigen-antiboy is very specific. On a microscope slide or in a test tube, a drop of red blood cells is mixed with a drop of serum containing a known antibody. There are two outcomes, i.e. the red blood cells clump (agglutination) or nothing happens. Figure 10-1 illustrates these two outcomes where A, B, and C are different magnifications. Agglutination is an indication of the correponding antigene's presence and no agglutination means that the antigen is absent.

During the time when Mendel's laws of inheritance were rediscovered, an Austrian physician Karl Landsteiner decided to inquire (in 1901) into whether lesser reactions might occur between the bloods of individuals from the same species. He took blood from himself and from five of his colleagues.

Figure 10-1

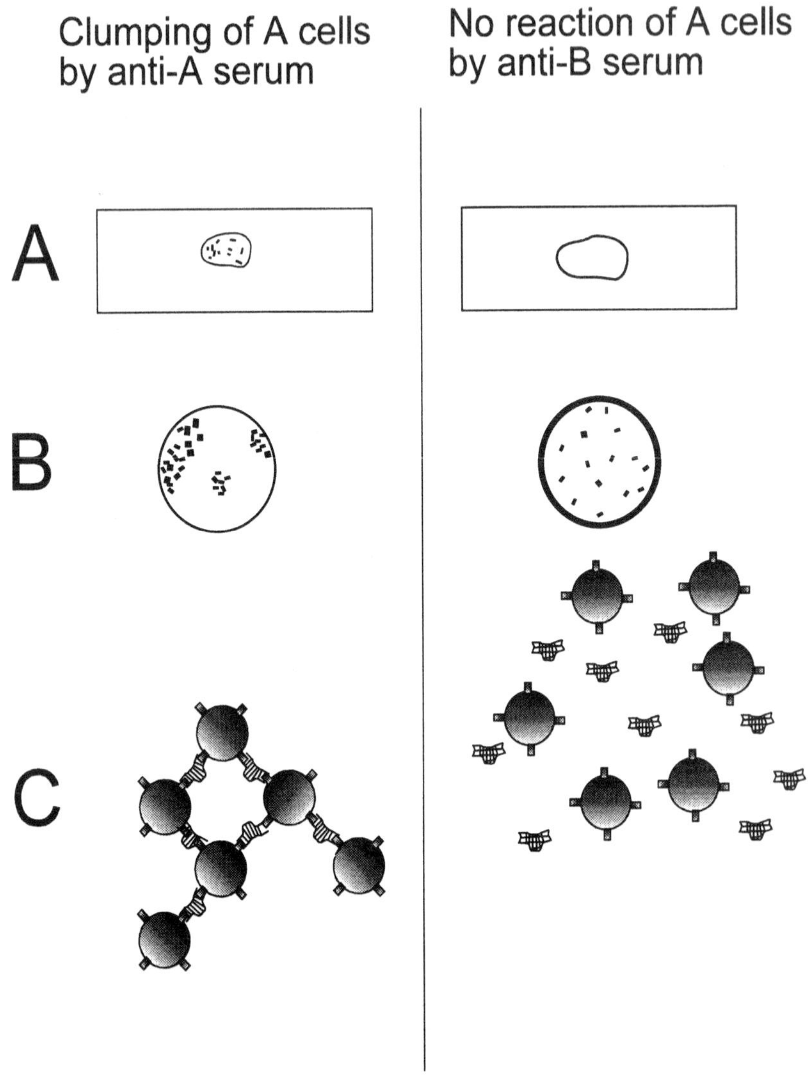
Clumping of A cells
by anti-A serum
No reaction of A cells
by anti-B serum
A
B
C

He then separated the red blood cells from the sera by centrifugation and subsequently recombined them in all possible pairings. Ho noted that the serum of Dr. Pletschnig agglutinated the cells of Dr. Sturli. However, Dr. Pletschnig's serum did not affect his own blood cells. The first letters of the alphabet were available to label the different sera and cells.

The alleles, *A*, *B*, and *O*, control the cell surface antigens of the *ABO* blood group system (Table 10-1).

ABO blood typing can be used to exclude an individual completely, however, since the majority of people have either *A* and/or *O* blood types, exclusions are hardly ever determined this way. On the other hand, an inclusion doesn't really provide a high probability of paternity, even if the rare *B* blood type was established between the alleged father and child. Using blood typing to determine paternity is not an easy chore for someone who does not understand genetics. This is because the *O* type refers to a lack of antigen, and is not determinable by standard blood typing methods. In other words, if someone is blood type *A*, they may actually be blood type *AO* or blood type *AA*. This

Table10-1

In vitro reaction of red blood with		Antigens on red blood cells	Blood type (phenotype)	Genotype
anti-A	anti-B			
+	0	A only	A	A/A or A/O
0	+	B only	B	B/B or B/O
+	+	Both A and B	AB	A/B
0	0	Neither A nor B	O	O/O

holds true for the *B* blood type as well, being either *BO* or *BB* as seen in (Tables 10-2 and 10-3).

Table10-2

		The Father is:				
		A	B	AB	O	
	A	A or O	A, B, AB, or O	A, B, or AB	A or O	
The Mother is:	B	A, B, AB, or O	B or O	A, B, or AB	B or O	The Child must be:
	AB	A, B, or AB	A, B, or AB	A, B, or AB	A or B	
	O	A or O	B or O	A or B	O	

Table10-3

		The Child is:				
		A	**B**	**AB**	**O**	
	A	A, B, AB, or O	B or AB	B or AB	A, B, or O	
	B	A or AB	A, B, AB, or O	A or AB	A, B, or O	
The Mother is:	***AB***	A, B, AB, or O	A, B, AB, or O	A, B, or AB	Ordinarily not possible.	The Fathe- must be:
	O	A or AB	B or AB	Ordinarily not possible.	A, B, or O	

HLA TYPING

The need to use organs for transplantation in humans has driven the knowledge of histocompatibility (histo in Greek means tissue). Rejection of an organ or tissue from the donor occurs if they posses histocompatibility antigens foreign to the recipient. These cell surface molecules (antigens) are polypeptides that combine with carbohydrates to form cell surface glycoproteins named Human Leukocyte Antigens (*HLA*). These molecules are coded for by the HLA genes.

HLA Typing is used to identify the HLA genes in a given individual. This can, like with *ABO* Typing, be used to determine paternity. The test is carried out either with cellular immunology techniques or with DNA procedures.

The *HLA* genes are divided into several different groups, the Class I genes *A*, *B* and *C*, and the Class II genes *DR*, *DQ* and *DP*. A given individual will have two copies of the several possible variants in each of these groups.

Chapter 11

PARENTAGE TESTING IN THE UNITED STATES: THE ROLE OF THE AMERICAN ASSOCIATION OF BLOOD BANKS*

By Robert W. Allen, Ph.D.

The involvement of the American Association of Blood Genetic Banks (AABB) in parentage testing can be traced to the testing to late 1970s and grew out of a need to establish paternity as a result of the increasing number of requests to AFDC (Aid to Families with Dependent Children) for support of children living in single-parent households. Claims had been steadily rising since the end of World War II, not due to death or disability, but rather due to abandonment of families by fathers and the rise in illegitimacy in the postwar period. As one means to try to contain costs, the Federal government passed the Title IV-D legislation in 1975, which established a mechanism for identifying fathers so that the responsibility for child support could be enforced upon them. Threshold levels of paternity establishment were set for states, that if met resulted in financial support of their programs; if the thresholds were not met, penalties resulted.

Genetic testing to establish paternity was first acknowledged officially by the legal profession with the passage of a law in New York in 1935 stating that ABO blood group test results could be introduced in court if performed

*Modified from an article published in *Profiles in DNA*, Vol. 2, No. 2. Courtesy of Promega Corporation.

by an individual of acceptable expertise. Expertise was defined as a doctoral degree and training in blood banking, hematology or genetics. In 1937 the MN blood group system markers were added to the list of acceptable markers. Additional systems were added as reagents and the understanding of the genetics of blood groups increased. In the mid-1970s, with the addition of HLA serology, paternity laboratories routinely became able to not only exclude falsely accused men but also to calculate probabilities of paternity for non excluded men that were compelling to the courts. Such powerful, yet complex, serological tests led both the American Bar Association and the American Medical Association to conclude that guidelines for performing tests and interpreting the results were needed. The first such fusion of law and science appeared in 1976 in the Family Law Quarterly. This collaborative work, prepared by physicians and legal experts, initiated a discussion on the qualifications needed for experts and for laboratories performing such tests, so test results could be accepted by the courts without question. The discussion even extended to proposing an accreditation program based on generally agreed upon standards.

DEVELOPMENT OF PARENTAGE TESTING STANDARDS

One of the authors of the paper, Harry Krause, JD, felt this topic to be so critical that he and others in the field organized a meeting of experts in parentage testing to discuss how best to perform such tests and interpret results. Underwritten by a grant from the Office of Child Support Enforcement, within the Department of Health and testing Human Services, this pivotal meeting was organized with the help of the AABB and was held in 1982 in Airlie, Virginia. In one of the opening lectures, Mr. Krause stated that the purpose of Airlie was to "arrive at a scientific consensus or, minimally, define what our disagreements are, in the service of the law".

He felt that Child Support Enforcement needed reform on two levels to reap maximum benefit from the Title IV-D legislation. First, a new framework for paternity actions was needed to improve quality and consistency of genetic testing, so as to expand the volume of claims that could often be processed by the courts. Secondly, the more efficient system must, nonetheless, provide fuller protection and safeguards for falsely accused men. It was clear that both goals could be realized if an accreditation program existed for parentage testing laboratories.

The AABB was a logical choice for involvement in parentage testing because of the reliance on serology for testing, an expertise largely found in blood bank laboratories, which of course were often associated with the AABB. The first official entry into parentage testing by the AABB came in 1978 when Dr. Byron Myhre, then President of the organization, recommended to the Board of Directors that a Committee on Parentage Testing beformed. Dr. Richard H. Walker assumed the role as Chair of the Committee, whose first charge was to assess the status of parentage testing in the United States, both in terms of the number of laboratories performing testing and the number of tests performed. The first survey performed by the committee indicated that in 1978 there were 259 laboratories performing about 11,000 tests, with a mean caseload of about 42 cases per year. By comparison, the Annual Report of 1995 and 1996 data published last year by the AABB indicated a total paternity test volume in the United States of approximately 170,000 cases, performed by 45 laboratories.

ACCREDITATION GUIDELINES

The first step in the formulation of standards by the Parentage Testing Committee was the publication of the Guidelines for Reporting Estimates of Probability of Paternity. No procedure or policy in this first publication was mandatory. Rather, practices in these guidelines were simply recommended. A publication of standards that included the words “must” and “shall” (i.e., requirements rather than recommendations) occurred with the implementation of an accreditation program in 1984. Accreditation and standard setting have undergone changes over the years, culminating in the present program, consisting of an on-site inspection every two years, based upon the recently published Fourth Edition of Standards for Parentage Testing Laboratories (1999). Other organizations engaged in genetic testing for identification have also established accreditation programs like that of the AABB, involving on-site inspection based upon a set of mandated standards and recommended practices.

The field of human identity testing is dynamic. Technologies are changing at a rapid rate and any accreditation program must likewise evolve to remain effective at providing quality assurance to the end user.

In the parentage testing area, one of the goals professed by Harry Krause, JD at the Airlie Conference in 1982 has been realized. The accreditation program of the AABB has succeeded in ensuring a high level of quality in laboratory testing in cases of disputed parentage. This consistent level of overall quality has resulted in general acceptance of test results from accredited laboratories by the courts without the need for expensive and time-consuming trials involving expert witnesses. It is likely that crime laboratories engaged in DNA typing will realize a similar benefit from a widespread accreditation program devoted to the forensic use of genetic testing.

SELECTED RESOURCES

1. Reinhard, J. (1999) Frozen in time. *National Geographics*, 196: 36-55.
2. Miale, J. et al. (1976) Joint AMAA/ABA guidelines: Present status of serologic testing in problems of disputed parentage. *Family Law Quarterly* 10, 247.
3. Annual Report on Parentage Testing for 1995 and 1996. American Association of Blood Banks.
4. Inclusion Probabilities in Parentage Testing (1983) R.H. Walker ed., American Association of Blood Banks, Bethesda, MD, p.xiv.
5. Foster, E.A., Jobling, M.A., Taylor, P.G., et al. (1998) Jefferson fathered slave's last child. *Science,* 396:27-28.
6. Lahn, B.T. and Page, D.C. (1999) Four evolutionary strata on the human X chromosome. *Science*, 286:964-967.
7. Jensen, M. (1998) All about Adam. *New Scientist*, July 35-39.
8. Kittles, R.A., Long, J.C., Bergen, A.W. et al. (1999) Cladistic association analysis of Y chromosome effects on alcohol dependence and related personality traits. *Proc. Natl. Acad. Sci. USA*, 96:4204-4209.
9. Sykes, B. and Irven, C. (2000) Surnames and the Y chromosome. *Am. J. Hum. Genet.*, 66:1417-1419.
10. Bradman, N. and Thomas, M. (1998) Why Y? *Science Spectra*, 14:32-37.
11. Hurles. M.E., Veitia, R., Arroyo, E. et al. (1999) Recent male-mediated gene flow over linguistic barrier in Iberia, suggested by analysis of a Y-chromosomal DNA polymorphism. *Am. J. Hum. Genet.*, 65:1437-1448.
12. Devaney, J.M., Girard, J.E., and Marino, M.A. (2000) DNA microsatellite analysis using ion-pair reversed-phase high-performance liquid chromatography. *Anal. Chem.*, 72:858-864.

13. Ventura, S.J., Mosher, W.D., Curtin, S.C., et al. (1999) Highlights in trends in pregnancies and pregnancy rates by outcome: estimates for the United States, 1976-96. *National Vital Statistics Reports*, 47(29):1-12.
14. Curtin, S.C. and Park, M.M. (1999) Trends in the attendant, place, and timing of births, and in the use of obstetric interventions: United States, 1989-97. *National Vital Statistics Reports*, 47(27):1-16.
15. Eisenbach, M. and Tur-Kaspa, I. (1999) Do human eggs attract spermatozoa? *BioEssays*, 21:203-210.

GLOSSARY

Alleles: Alternate forms of the same gene. When a genetic locus is polymorphic, there are alleles for that gene. Example are A, B, and O of the ABO blood gene.

Alleged Father (AF): Putative Father (PF): A man who is contesting an accusation of paternity or wishes to establish it when a question of paternity exists.

Alzheimer's disease: a disease that causes memory loss, personality changes, dementia and, ultimately death. Not all cases are inherited, but genes have been found for familial forms of Alzheimer's disease.

Amino acid: Any of a class of 20 molecules that combine to form proteins in living things.

Antigen: Genetic product (protein) that is detected by immunological techniques (using antibodies).

Acquired mutations: Gene changes that arise within individual cells and accumulate throughout a person's lifetime; also called somatic mutations (see Hereditary mutation).

Autoradiogram, Autoradiograph (Autorad): The visual recording on x-ray film that displays polymorphic DNA fragments present in the mother, child, and alleged father.

Autozome: Any of the non-sex determining chromosomes. Human cells have 22 pairs of autozomes.

Band: The visual image on an autoradiograph that represents a particular DNA fragment.

Bases: (In DNA) Four building blocks of nucleic acids. They are Adenine (A), Thymine (T), Guanine (G), and Cytosine (C). They are complementary in that the only stable base pairs are AT/TA and GC/CG. The length of DNA fragments is measured in base pairs (bp).

Base pairs: The two complementary, nitrogen rich molecules (bases) held together by weak chemical bonds. Two strands of DNBA are held together in a shape of a double helix by the bonds between their base pairs.

Bayes' theorem (as applicable to parentage testing): A statistical procedure to assess the relative probability of two alternative possibilities based on acquired information.
W = (CPI) (Pr) (CPI) (Pr) + (1 - Pr)
W = Probability of Paternity, CPI = Combined Paternity Index, Pr = Prior Probability

Biological Father: The man who provided the paternal genetic material of a child.

Carrier: A person who has a recessive mutated gene, together with its normal allele. Carriers do not usually develop disease but can pass the mutated gene on to their children.

Carrier testing: testing to identify individuals who carry disease-causing recessive genes that could be inherited by their children. Carrier testing is designed for healthy people who have no symptoms of disease, but who are known to be at high risk because of family history.

Cell: Small, watery, membrane-bound compartment filled with chemicals; the basic subunit of any living thing.

Chromosome: A thread like structure found in the nucleus of the cell that carries genetic information arranged in a linear sequence. It consists of a complex of nucleic acids and proteins. At certain times during cell life chromosomes appear as distinct units. In humans there are 46 in the nucleus of every nucleated cell arranged in 23 pairs.

Clone: A group of identical genes, cells, or organisms derived from a single ancestor.

Codominant: Two alleles that are both seen in one individual. Example: AB blood type (see dominant and recessive).

Combined Paternity Index (CPI): The product of the individual PI values calculated for each independent test. The CPI is the ratio based on the testing of the probability that the alleged father is the biological father (x) to the probability that a random man is the biological father (y). This value is calculated when the alleged father can not be excluded on a minimum of 2 tests.

DNA (Deoxyribonucleic Acid): The molecule that encodes genetic information. DNA is a double-stranded helix held together by weak bonds between base pairs of nucleotides. DNA is a genetic material that is routinely tested in parentage evaluations.

Diploid: Containing a complete set (one from each parent) of homologous (similar) chromosomes.

Direct Exclusion: When a characteristic is found in the child that is not in the child's mother, it must have come from the biological father. If an alleged father does not have the factor, he is excluded for that system. If an alleged father is excluded on two independent tests, he is excluded as the biological father of the child.

Dominant: When on allele's expression in an individual masks the expression of another allele in that individual. Example: people with type A blood can either have two A alleles or one A allele and one O allele. The O allele will not be seen due to the presence of the dominant A allele (see recessive and codominant).

Double helix: The structure of DNA. Two strands of DNA coiled around each other forming a spiral staircase like structure.

Electrophoresis: The method by which biological molecules are separated according to certain characteristics. In DNA parentage testing the negatively charged DNA fragments are cut and separated by size in a porous material such as agarose, by application of an electric field.

Gene: The fundamental unit of heredity; an ordered sequence of nucleotide base pairs to which a specific product or function can be assigned. Many genes have alternate forms referred to as alleles (see allele).

Genotype: The actual alleles present at a given gene locus. Example: individuals who have type AB blood have the genotype AB, a person who has type A blood could have the Genotype AA or AO.

Genetic Marker: An inherited (genetic) characteristic that can be recognized in a tested individual and is used in differentiating genetically related individuals from nonrelated individuals.

Haploid: Containing only 1 of each pair of the 46 chromosomes (seen in eggs and sperm).

Hardy-Weinberg Law: The mathematical expression of the equilibrium of allelic genetic markers within the population. A mathematical model that allows for the prediction of occurrence of genetic markers in a population and states that in a large random intra-breeding population, not subjective to excessive selection or mutation, the gene and genotype frequency will remain constant over time.

$$(p + q)^2 = p^2 + 2pq + q^2 = 1$$

whereas p and q represent the frequency of two allelic genes in a two allele system, p^2 and q^2 represent the frequencies of homozygotes, 2pq represents the frequency of heterozygotes. In a multiple allelic system: $(p + q + r + s)^2 = 1$, whereas r and s represent additional alleles in a multiple allele system.

Heterozygote (heterozygous): Individual who has two different forms (alleles) of a specific gene. Example: people with the AB blood type have two alleles (A and B) for the blood type gene.

Homologous: In chromosomes, a pair of chromosomes that contain the same gene (although perhaps different alleles). There are 23 pairs of homologous chromosomes in humans; one of each was contributed by each parent.

Homozygote (homozygous): Individual who has identical forms (alleles) of a given gene. Example: people with the O blood type have two O alleles of the blood type gene.

Indirect Exclusion: A presumed exclusion based on the observation that the child and AF are homozygous for different alleles. It is possible that the child and AF actually share a null gene that is not detected. An indirect exclusion is not considered 100% conclusive, and usually before an opinion of nonpaternity can be rendered, a search is made for a second direct or indirect exclusion.

Kilobase: One thousand base pairs of DNA. Often abbreviated as kb.

Mean Probability of Exclusion: The probability that a single genetic system will exclude a non-father from paternity prior to testing, i.e., independent of the blood types of the trio tested.

Microsatellites: short sequences of nucleotides (example: ATCG) which are repeated over and over again a number of times in tandem.

Minisatellites: longer sequences of nucleotides (example: ATCG) which are repeated over and over again a number of times in tandem.

Mutation: Any change in the sequence of genomic DNA.

Paternity Index (Likelihood Ratio): For an individual genetic system it is the ratio between the chance that the alleged father may pass the paternal gene compared to the chance that a random man may pass the paternal gene to the child. It is sometimes referred to as the genetic odds in favor of paternity, given the genetic findings in the mother-child-alleged father.

Phenotype: The measurable genetic markers in a given individual at a specific locus. It may or may not reflect the actual alleles present (see genotype). Example: a person with type A blood is said to have the A phenotype but could have the AO genotype.

Polymorphism: The existence of more than one form of a genetic trait.

Population Database: A random sampling of alleles for a given probe/racial group combination used to determine the frequency of an allele in the population.

Population Genetics: A discipline that examines the character and extent of inherited differences within and between populations.

Power of Exclusion (A): The ability of a genetic test to exclude a falsely accused man of paternity and is dependent upon the phenotypes of the mother and child and the ethnic background of the alleged father.

$$A = (1 - p)^2$$

where p = the frequency of the paternal allele in the child.

Prior Probability (Pr): An assigned numerical value of from 0 - 1 (ranging from impossible to certain) of the non-genetic evidence used in evaluating paternity. It is estimated on the basis of the circumstances surrounding the event, e.g. casual aquaintance versus an intimate relationship. A prior probability of 0.5 is considered neutral and is an equal weighting of the nongenetic evidence for and against paternity.

Probability of Paternity (W): A statistical means of describing the likelihood that the alleged father is the biological father of the child. This value incorporates the genetic (Paternity Index) and nongenetic (Prior Probability) information.

Probe: Short segment of DNA that is tagged with a radioactive or chemical tracer and is then used to detect the presence of a particular DNA sequence through hybridization to its complementary sequence.

Random Man: A theoretical man of unknown genetic comparison from a population with the same racial background as the alleged father.

Random Man Not Excluded (RMNE): The frequency with which men selected at random from the same racial group as the alleged father and tested to the same extent as the alleged father would not be excluded as the biological father of the child.

Recessive: A trait that is partially or completely masked by the effect of a dominant allele. The dominant A or B allele masks the presence of the recessive O allele.

Red Cell Antigens: Genetic markers found on the surface of red cells, e.g. ABO, Rh, MNSs.

Restriction Enzyme: A bacterial enzyme, which recognizes a specific set of bases and the DNA at that specific site.

Restriction Fragment: A piece of DNA identified by a specific restriction enzyme.

Restriction Fragment Length Polymorphism (RFLP): Variations in the size of DNA fragments produced by a restriction endonuclease at a polymorphic locus.

Single Locus Probe: A DNA probe that detects a single DNA region creating one ("homozygote") to two (heterozygote) bands on an autoradiograph.

SNPs: single nucleotide polymorphisms in which a particular nucleotide is changed (like A is changed to G). Since SNPs and indels (stable alus) are very rare, they also are known as unique event polymorphisms (UEPs).

Variable Number of Tandem Repeats (VNRT): DNA locations that are comprised of short (i.e. 10-70 base pairs) repeats arranged in tandem. The length of the location will depend on the number of times the sequence is repeated. The number of repeats will vary among non-related individuals.

Appendix I

A LEGAL PERSPECTIVE - NOTES FROM AN ATTORNEY'S OFFICE

By James P. Reape

New federal law

Effective 1996, the federal government amended 42 U.S.C. §666(a)(5) to require all states to have statutorily proscribed procedures to improve effectiveness of child support enforcement, to provide for and to insist on genetic testing in contesting paternity cases.

The child and all other parties are required to submit to genetic testing upon the request of any party. That party must support her request with a sworn statement alleging paternity, and setting forth facts establishing a reasonable possibility of the requisite sexual contact between the parties or denying paternity and setting forth facts establishing a reasonable possibility of the nonexistence of sexual contact between the parties.

The statute also allows for the admissibility of genetic testing results if the test is of a type generally acknowledged as reliable by accreditation bodies designated by the Secretary and performed by a laboratory approved by such an accreditation body. **The genetic test results are admissible as evidence of paternity without the need for foundation testimony or other proof of authenticity or accuracy, unless objection is made.**

The federal law allows a state to establish a rebuttable or conclusive presumption of paternity upon genetic testing results, which indicate a threshold probability that the alleged father is the father of the child.

The paternity decision may be made by a judicial or administrative body if shown by clear and convincing evidence of paternity.

<u>The party has no right to jury trial.</u>

In more detail, 42 USC §466 (5) (A) - (M) provides for the following:

The state must have procedures, which permit the establishment of paternity of a child at any time from birth until the child turns 18. The procedures shall apply for a child for whom paternity has not yet been established or whom a paternity action was brought but dismissed because a SOL of less than 18 years was then in effect in the State.

The State is required in a contested paternity case to require the child and all other parties to submit to genetic tests upon the request of any such party, if the request is supported by a sworn statement by the party alleging or denying paternity, with facts establishing reasonable possibility of the requisite or non-existence of sexual contact between the parties.

The state agency must pay the costs of such genetic test, subject to recoupement from alleged father if paternity is established and to obtain additional testing in any case if an original test is contested, upon request and advance payment by the contestant.

A mother and a putative father can voluntarily acknowledge paternity through a simple civil process by signing a Voluntary Paternity Acknowledgment . The mother and father must be given notice orally and in writing of the legal consequences of the rights and responsibilities that arise from signing the acknowledgment.

Procedures must include a hospital based program for the voluntary acknowledgment of paternity focusing on the period immediately before and after the birth of the child. The State agency responsible for maintaining birth records must offer voluntary paternity establishment services.

The state admit into evidence, for purposes of establishing paternity, the results of any genetic test that is generally acknowledged as reliable by accreditation bodies designated by he Secretary and performed by a laboratory approved by such an accreditation body. An objection to genetic testing results must be made in writing not later than a specified number of days before any hearing at which the results may be introduced into evidence and the test results must be admissible as evidence of paternity without the need for foundation testimony or other proof of authenticity or accuracy, unless objection is made.

The state must have procedures which create a rebuttable or at the state's option, a conclusive presumption of paternity upon genetic testing results indicating a threshold of probability that the alleged father is he father of the child.

Parties to an action to establish paternity are not entitled to a jury trial.

Temporary orders may be issued upon a motion by a party requiring child support pending an administrative or judicial determination of parentage, if there is clear and convincing evidence of paternity on the basis of genetic tests or other evidence

Putative fathers must have a reasonable opportunity to initiate a paternity action.

***SURVEY CHART OF STATES AND THEIR STATUTES**

Connecticut is the only state that specifically limits admissibility of genetic test evidence to instances in which the defendant is excluded as the father (Moore v. McNamara, 201 Conn. 16, 513 A.2d 660 (1986)).

Arizona and Pennsylvania statutes are ambiguous in that wording of the statue does not make clear whether the evidence is exclusionary or inclusionary.

All other states are inclusionary - the genetic test evidence may be admitted if the results include the putative father in the group of possible fathers.

Some states have a statute creating a Presumption of Paternity if the genetic test results report a probability of paternity greater than or equal to a specified percentage. Alaska is 95% rebuttable by clear and convincing evidence, California is 99% rebuttable by preponderance of the evidence (with a paternity index of 100), Colorado is 97% the degree of proof required to rebut is not specified, Florida is 95%rebuttable by a preponderance of the evidence, Main is 97% rebuttable by clear and convincing evidence , Oklahoma 95% rebuttable by clear and convincing evidence, Wisconsin 99%the degree of proof required to rebut is not specified. "Admissibility of Genetic Testing in Paternity Litigation: a Survey of State Statutes" by D. H. Kaye, Ronald Kanwischer, Family Law Quarterly, Volume XXII, Number 1, Spring 1988

UNIFORM ACT ON BLOOD TESTS TO DETERMINE PATERNITY

The following states contain this language in their statutes: California, New Hampshire, Oregon, and Pennsylvania

UNIFORM ACT ON PATERNITY

Adopted in Kentucky, Maine, Mississippi, New Hampshire, Rhode Island, and Utah

UNIFORM PATERNITY ACT (1973)

Adopted in Alabama, California, Colorado, Delaware, Hawaii, Ilinois, Kansas, Minnesota, Missouri, Montana, Nevada, New Jersey, New Mexico, North Dakota, Ohio, Rhode Island, Washington, and Wyoming.

The National Conference of Commissioners on Uniform State Laws promulgated the Uniform Paternity Act in 1973, at a time when states needed new legislation on this subject because the bulk of current law on the subject of children born out of wedlock was either unconstitutional or subject to grave constitutional doubt.

The United States Supreme Court, under Equal Protection Clause of 14th Amendment of US Constitution, mandated equal legal treatment of legitimate and illegitimate children in broad range of substantive areas:

Weber v. Aetna Casualty & Surety Company, 92 S.Ct. 1400, 1406-07 (1972) "...no child is responsible for his birth and penalizing the illegitimate child is an ineffectual as well as an unjust way of deterring the parent. Courts are powerless to prevent the social opprobrium suffered by these hapless children, but the Equal Protection Clause does enable us to strike down discriminatory laws relating to status of birth where - as in this case - the classification is justified by no legitimate state interest, compelling or otherwise." (facts - workmen's compensation benefits related to the death of their father are due dependent, unacknowledged, illegitimate children)

Gomez v. Perez, 93 S.Ct. 872 (1973) " We have held that under the Equal Protection Clause of the Fourteenth Amendment a State may not create a right of action in favor of children for the wrongful death of a parent and exclude illegitimate children from the benefit of such a right. ....We therefore hold that once a State posits a judicially enforceable right on behalf of children to needed support from their natural fathers there is no constitutionally sufficient justification for denying such an essential right to a child simply because her natural father has not married her mother. For to do so is 'illogical and unjust." We recognize the lurking problems with respect to proof to paternity. Those problems are not to be lightly brushed aside, but neither can they be made into an impenetrable barrier that works to shield otherwise invidious discrimination" (facts - the illegitimate child is guaranteed the right of support from his father).

In providing substantive legal equality for all children regardless of the marital status of their parents, the present Act merely fulfills the mandate of the Constitution.

To identify the father, the Act first sets up a network of presumptions which cover cases in which proof of external circumstances (marriage between the mother and a man) indicate a particular man to be the probable father. All presumptions of paternity are rebuttable in appropriate circumstances.

The next major function of the Act is the ascertainment of paternity when no external circumstances presumptively point to a particular man as the father.

It is expected that this Act will fulfill an important social need in terms of improving the state's systems of support enforcement.

Since the 1973 promulgation of this Act by the Conference, the U.S. Supreme Court decided several cases involving the rights of unwed fathers. Conference then promulgated the Uniform Putative and Unknown Fathers in 1988.

SECTION 1 Parent and Child Relationship is the legal relationship existing between a child and his natural or adoptive parents incident to which the law confers or imposes rights, privileges, duties and obligations.

SECTION 2 The parent and child relationship extends equally to every child and to every parent, regardless of marital status of parents. (SECTION 1&2 establish the principle that regardless of the marital status of the parents, all children and all parents have equal rights with respect to each other).

SECTION 3 Parent and child relationship between child and natural mother may be established by proof of her having given birth of the child, or under this Act; the natural father may be established under this Act; an adoptive parent may be established by proof of adoption or under the Revised Uniform Adoption Act.

SECTION 4 **A man is presumed to be the natural father of a child if:**

1. he and the child's mother are or have been married to each other and the child is born during the marriage or within 300 days after the marriage is terminated by death, annulment, declaration of invalidity, or divorce, or after a decree of separation is entered by a court;
2. before the child's birth, he and the child's natural mother have attempted to marry each other; if the attempted marriage could be declared invalid only by a court, the child is born during the attempted marriage or within 300 days after its termination by death, annulment, declaration of invalidity or divorce; if the attempted marriage is invalid without a court order, the child is born within 300 days after the termination of cohabitation;
3. after the child's birth, he and the child's natural mother have married or attempted to marry, although the attempted marriages or could be declared invalid; he has acknowledged his paternity of the child in writing filed with the appropriate court or Vital Statistics Bureau; with his consent, he is the named father on the child's birth certificate; he is obligated to support the child under a written voluntary promise or by court order;
4. while the child is under the age of majority, he receives the child into his home and openly holds out the child as his natural child;
5. he acknowledges his paternity of the child in a writing filed with the appropriate court or Vital Statistics Bureau, which shall promptly inform the mother of the filing of the acknowledgment, and she does not dispute the acknowledgment within a reasonable time after being informed thereof, in a writing filed with the appropriate court. If another man is presumed under this section to be the child's father, acknowledgment may be effected only with the written consent of the

presumed father or after the presumption has been rebutted.

(B) A presumption under this section may be rebutted in an appropriate action only be clear and convincing evidence. If two or more presumptions arise which conflict each other, the presumption, which on the facts is founded on the weightier considerations of policy and logic controls. The presumption is rebutted by a court decree establishing paternity of the child by another man.

SECTION 5 Artificial Insemination

SECTION 6 Determination of Father and Child Relationship; Who May Bring Action, When Action May be brought

(A) A child, his natural mother, or a man presumed to be his father under 1, 2, 3 may bring action

(1) at any time for purpose of declaring existence of the father and child relationship presumed under 1, 2, 3

(2) for purpose of declaring non-existence of the father and child relationship presumed under 1, 2, 3 only if the action is brought within reasonable time after obtaining knowledge of relevant facts, but no later than 5 years after childs birth. After presumption has been rebutted, paterntiy of the child by antoher man may be determined in the same action if he has been made a party

(b) any interested party may bring an action at any time for the purpose of determining existence or non-existence of the father and child relationship presumed under 4 or 5

(c) an action to determine the existence of the father and child relationship with respect to a child who has no presumed father may be brought by the child, the mother or personal representative of the child, the appropriate state agency, the personal representative or parent of the mother if the mother has died, a man alleged or alleging himself to be the father or the personal representative or a parent of the alleged father if the alleged father has died or is a minor

(d) Regardless of its terms, an agreement, other than t he approved by the court, between alleged or presumed father and the mother or child, does not bar an action under this section.
If action under this section brought before birth of child, proceedings shall be stayed until after the birth.

SECTION 7 Statute of Limitations

An action to determine existence of father and child relationship as to child who has no presumed father may not be brought later than 3 years after birth of child. Action brought by or on behalf of a child whose paternity has not been determined is not barred until 3 years after the child reaches age of majority. (U.S. Supreme Court decisions speak in terms of child's substantive right to a legal relationship with his father, unreasonable to bar the child's action by reason of another person's failure to bring paternity action at an earlier time).

<u>*CONSTITUTIONALITY OF TESTING</u>

The law ought to protect the interests of individuals in having their genetic information kept private. New testing technologies pose serious risks to freedom and privacy. They risk increasing the power that a very few people will hold over many. Christopher Blakely, Scientific Testing and Proof of Paternity: Some Controversy and Key Issues for Family Law Counsel,
57 La.L. Re. 379-438

<u>In re J.M.</u> (1991) 590 So.2d 565. Although an alleged father has a right of privacy and to be free from unreasonable searches and seizures, those rights are not absolute and may be reasonably regulated when the state has a sufficiently weighty interest.

<u>In re J.M.</u> (1991) 590 So.2d 565. Blood and tissue testing is a search and seizure however, the significant state interest in the welfare of children and the conservation of public assistance funds justifies the intrusion. The compelling interest stems from the pervasive concern for the welfare of children. A blood test is minimally intrusive, relatively painless, and medically safe. In facilitating a determination of paternity, blood tests are highly reliable and unequaled in evidentiary value.

<u>Matthews v. Eldridge</u> (1976) 424 U.S. 319. The United States Supreme Court applied a balancing test to determine what procedures are constitutionally required to protect the alleged father's rights. The defendants privacy interests were balanced against the state's compelling interest in protecting its children.

<u>Michael H. V. Gerald D.</u> (1989) 491 U.S. 110. The conclusive marital presumption is constitutional and does not deprive the natural father of his due process right to establish paternity.

<u>Dawn D. V. Superior Court</u> (1998) 17 Cal.4th 932. An alleged biological father does not have a Constitutionally protected due process right to establish a parental relationship with a child born to his former lover and her husband during the marriage but while the lover and husband were not cohabiting. Summary judgment was granted to the mother and her husband.

<u>Rodney F. V. Karen M.</u> (1998) 61 Cal.App.4th 1311d. The

conclusive presumption was applicable and the court should not order DNA paternity testing for a man who is not the presumed father and husband and wife were cohabiting at the time of conception.

Michael M. V. Giovanna F. (1992) 5 Cal.App.4th 1272. Where the biological father takes all the possible steps to establish a relationship with the child, he has standing to bring a paternity action where the child has a presumed father in an intact family, but there must be a balancing of interests.

Procedural Due Process requires that the party alleging paternity make a preliminary showing that there is a reasonable possibility of paternity. "An individual 's constitutional right to due process is implicated when compulsory blood testing is ordered by a court, Breithaupt v. Abram (1957) 352 U.S. 432.

Generally the laws prohibit obtaining genetic information from and individual and taking an individuals tissue sample for DNA testing proposed without first obtaining informed consent. Exceptions to these protections include the identification of a deceased person where the government is authorized by a specific law such as for criminal investigations, screening newborns, anonymous research and for the establishment of paternity. In these cases, consent is not necessary. Sudwischer v. Hoffpauir, 589 So.2d 474

STANDARD OF PROOF

Rivira v. Minnich (1987) 483 U.S. 574. Affirmed the constitutionality of applying the preponderance of evidence standard to establish paternity.

Most states call for proof by clear and convincing evidence when the alleged father is deceased. Tennessee requires clear and convincing evidence after alleged father's death - and paternity must be established before the property of the deceased father vests in persons other than the claimant. Bilbrey v. Smithers, 1996 WL 4949900

RIGHT TO JURY TRIAL see fed law, no right to jury trial. Decisions can be made by courts or administrative bodies.

STANDING see fed law, mother, child, alleged father, presumed father (not disinterested parties)

RIGHT TO ATTORNEY see fed law, only required for indigent fathers

***EXCLUDING VS. ESTABLISHING PATERNITY**

Paternity can be established in three ways -

1. Court ordered - The state can order a paternity test if Public Assistance is involved. The mother is requested to inform the office who the father is so that a child support order can be ordered, and the public assistance can be reimbursed to the state.
2. Voluntaration of Declaration of Paternity (Provided to the unwed mother and father at the time of birth and before release from the

hospital)

3. Civil Action - a private party can institute a paternity determination to establish rights to support, custody, visitation, and inheritance rights.

***THE MARITAL PRESUMPTION**

"The husband of a mother is the legal father of her child if the two were married at the time of conception or birth."

In many states where this is a conclusive presumption, which means that even if another man can prove through DNA or blood testing that he is the child's father, the court will not even consider his petition to establish paternity or order blood tests or hear evidence. There are strong policy reasons to protect the marital family unit. This strong presumption is in favor of legitimacy for the child, who in the past were not entitled to support or inheritance and subject to social disdain because of illegitimacy.

In many states this is a rebuttable presumption, but rebuttable only by clear and convincing evidence. Some states allow only the mother or her husband to bring an action seeking to rebut the marital presumption and not the putative father. Some courts will only order blood tests if it is in the child's best interests, giving strong weight to whether the child is already in an intact family. If mother and her husband unit in opposition, the court will usually refuse to order blood tests. The court will also look to see if the biological father has established a relationship with the child. Most courts will not allow an unwed father to use the marital presumption as a defense against a paternity action.

MARITAL PRESUMPTION CASES

Michael H. V. Gerald D. (1989) 491 U.S. 110, The conclusive presumption of California Family Code §7540 and 7541 is constitutional and does not deprive the natural father of the due process right to establish paternity.

Adoption of Michael H. (1995) 10 Cal.4th 1043. To achieve presumed father status, the biological father must promptly come forward and demonstrate full commitment to his parental responsibilities once he know or reasonably should know of pregnancy.

Rodney F. V. Karen M. (1998) 61 Cal.App.4th 1311d. The conclusive presumption was held applicable and the court should not order DNA paternity testing for a man who is not the presumed father and the husband and wife were cohabitating at the time of conception).

Steven W. V. Matthew S. (1995) 33 Cal.App.4th 1108. The conclusive presumption of paternity was inapplicable were the husband and wife were not cohabiting on the date of conception, a weekend tryst did not establish cohabitation.

Alicia R. V. Timothy M (1994) 29 Cal.App.4th 1232. The conclusive presumption will not be applied where policies underlying the presumption would not be served and a ridiculous result would ensue.

County of Orange v. Leslie B. (1993) 14 Cal.App.4th 976. The conclusive presumption will not be applied where policies underlying the presumption would not be served and a ridiculous result would ensue. The conclusive presumption was inapplicable where the marital unit ceased to exist before the birth of the child and the child knew husband was not her father.

***SCIENTIFIC TESTING**

Proving Paternity can be done by blood testing or DNA testing. The blood testing method currently in favor is the Human Leucocyte Antigen (HLA) test. White blood cells are examined and can exclude a man as the father as close to 99% certainty. This evidence is viewed as conclusive evidence denying paternity, it is not done to establish paternity. In most states, even with evidence of 90-99% paternity, creates only a rebuttable presumption of paternity. The alleged father can still produce evidence to rebut the presumption, such as impotency, sterility or not having asexual relations with the mother. Then that evidence is weighed against the blood tests.

DNA testing is the most recent typed used and the most accurate. The DNA makes a pattern like a bar code, each person's having a unique DNA. The child's DNA "fingerprint" is compared with the DNA fingerprints of the mother and putative father and then it is ascertained whether the man is the biological father. DNA testing is more accurate than HLA testing in that it yields a higher degree of accuracy. DNA testing is more expensive than blood grouping. Courts have just begun to consider whether or not to admit it into evidence within the last ten years.

Scientific testing should not be the sole factor in determining paternity. Alone, scientific testing is insufficient to meet the standard of proof to prove paternity. McKenzie v. Thomas, 1996 WL 375017. Corroboration still should be required because DNA testing is challengeable due to procedural errors which can occur at each stage of the procedure - errors in taking the samples, storage, shipping, testing, breach of or failure to verify chain of custody in proper sworn affidavit. To attack DNA testing rests solely on finding errors. In a case which the evidence as a whole was unfavorable to the plaintiff's case, the fact that the blood or tissue tests show a high probability that the defendant was the biological father was not allowed without other credible evidence (such as sexual access during the critical period) to establish paternity by preponderance of evidence. Suire v. Robison, 511 So.2d 1361

Some states however provide that when a person attempts to challenge a support order on the basis of non paternity without externally obtained clear medical proof, the challenge would be rejected. Leiter v. Scott, 654 N.E.2d 742

"When we allow scientists or statisticians to usurp the fact-finding role, we destroy the value, indeed the pertinence of the judicial system" "If DNA tests are not performed properly, you cannot distinguish one person from another, statistically or otherwise." Gerald Sheindlin, DNA: Is the Presentation of Statistical Evidence Necessary? 214 N.Y. L.J. (Aug. 4, 1995)

THE PATERNITY INDEX

The Paternity Index is " The probability that a cross between the defendant and the mother would produce an offspring with the child's phenotypes and the corresponding probability for a random selection of genes from the male population." D.H. Kaye, Plemel as a Primer on Proving Paternity, 24 Williamette L. Rev. 867, 877 (1988)

First step - determine the prior odds that the defendant is the child's father. Experts assume that the prior odds are 1:1, assuming that either the plaintiff or the defendant is or is not telling the truth.

Second step - multiply the odds of 1:1 by a fraction, numerator of which is the probability that the defendant is the child's father and the denominator is the probability that a man chosen at random from the population is the father. (Numerator is the probability in terms of percentage that a man having the d's genotypes would transmit the necessary genes to the child, given the mothers genotypes. Denominator is the probability in terms of percentage, that a mythical random ly chosen man would transmit those genes to t he child, taking into the account the mother's genes. When that fraction is computed into a whole number, that number, expressed as a percentage, is the Paternity Index. It is a comparison between the probability that the defendant transmitted the necessary genes to the child with the statistical incidence of those genes in the general population. Homer H. Clark, Jr. & Carol Glowinsky, Domestic Relations: Cases & Materials 328 (5th ed. 1995)

DNA testing in a paternity action is a six step process to determine whether the molecular structure of the alleged father matches that of his alleged child so that a scientist could say that the man was the father, or if the prints do not match to say the man is not the father. The alleged father's DNA is compared to that represented in a laboratory database containing samples from at least 100 men of the same or similar race, calculating the frequency with which the subject's fingerprint or other bodily material is found in such a population. From this calculation the likelihood of fatherhood is supposedly determined. David B. Jackson, DNA Fingerprinting and Proof of Paternity, 15 Fam. L. Rptr. 3007 (1989)

*CONSEQUENCES OF PATERNITY DETERMINATION
RIGHTS AND RESPONSIBILITIES

The parent and child relationship is established. With that comes the duty to support the child and to provide for its general welfare, including medical and education responsibilities.

*STATUTE OF LIMITATIONS

Most states have adopted a Statute of limitation on paternal determination for the purposes of support. The statute of limitations varies from state to state and ranges from two years after the birth of the child, to a "reasonable time".

The Supreme Court however has held that even a long statute of limitations is in violation of the equal protection clause of the 14th Amendment, finding that the illegitimate child has the same rights to the resources of his father as the child born in wedlock. The Supreme Court also held that there is no State interest in having a statute of limitations. In paternity testing, blood testing and DNA testing evidence is not subject to deterioration, which is the main reason for a statue of limitations. "Statute of Limitations in Paternity Proceedings: Barring an Illegitimate's Right to Support" (1983) 32 ALR 567

Under the Uniform Parentage Act, paternity must be established by the time the child reaches the age of majority for support purposes. However, under the Uniform Probate Code there is no strict time limitation to determine heirship of an intestate. This is to ensure that a potential heir is not prevented from asserting the right to have his or her status as an heir adjudicated.

Where states have not adopted the Uniform Probate Code, under the intestacy statutes of these states, a child can establish paternity by adjudication before the putative father's death or after the putative father's death by presenting clear and convincing evident that the man was the father. Bilbrey v. Smithers (1996) 937 S.W.2d 803

In re Heirship of McCleod (1987) 506 So.2d 289, the court held that the statutes providing an action to determine heirship do not prescribe any time limit. The action is needed only where there is a question about an heirship and no statute runs until the question is raised.

Little v. Smith (1997) 943 S.W.2d 414. Established a four-year limitation period for heirship determination where the estate had been closed, involving a claim by an adopted child against the closed estate of her biological grandmother. Discussed was the public policy argument that the finality of estates outweighs an individuals rights to establish heirship.

***EXHUMATION TO DETERMINE PATERNITY**

Increased paternity testing has been the predictable result of the exploding DNA phenomenon and accounts for a surprising increase in requests for exhumation of the putative father's remains to administer such testing. Ralph C. Brashier, Children and Inheritance In the Nontraditional Family, 1996 Utah L. Rev. 93.

Most courts will consider an exhumation request if there is reasonable belief that evidence to be found is probative to determine paternity and if there is sufficient evidence that the bone or tissue is retrievable. In Batcheldor v. Boyd (1993) 423 S.E.2d 810, the court ordered an exhumation to determine paternity. The information sought was reasonably calculated to lead to admissible evidence and the defendant had shown good cause to support exhumation of the body.

In Wawrykow v. Simonich, (1994) 652 A.2d 843, the court ruled exhumation permissible when reasonable was show, all evidence presented leads the court to believe that the child's paternity will be discovered upon exhumation. The court must find that the availability of blood or tissue samples of the decedent is not compromised by the passage of time and the embalming, rendering their retrieval improbable or unlikely to produce results from.

However if the DNA of the decedent is non-existence or cannot be achieved, the court may order a blood sample from a relative of the deceased. In Sudwischer v. Estate of Hoffpauir, (1991) 589 So.2d 474, a putative daughter sought determination whether the intestate was her biological father and the court ordered the legitimate daughter to submit or blood testing for DNA comparison purposes.

Appendix II

DNA TESTING RESOURCES FOR YOUR FAMILY

INDEX

ABO 113
Aid to Families with Dependent Children 119
Allele
  Dominant 34
  Recessive 34
  Segregation 34
Amelogenin 77

Barger, Herbet 100
base pairs 40
bases 39
Basques 98
Blood typing 113
brachydactyly 35

Capillary electrophoresis 64
Carr, John 99
Catalans 98
Chain of Custody 27
Chromosomes 43
Combined Paternity Index 87
Crick, Francis 32

Determination of Father and Child Relationship 140
Distant Family Relationship 101
DNA
  Biological sources 48
  Genomic 44
  Mitochondrial 46
  Profiling 67
DNA Polymerase 55

Electrophoresis 61
  Capillary 64
  Gel 61
Exclusion 86
Exhumation to determine paternity 147

Faerman, Marina 78
Foster, Eugene 99

Gels 63
gene 40
Genesis 96
Goldman, David 100
Grand-Parentage 93

HaeIII 69
Hammer, Michael 96
Hemings, Sally 99
Heterozygous 81
Hillblom, Larry 48
Homozyhous 81

International Society for Forensic Haemogenetics 74

Jefferson, Thomas 99
Johann, Gregor Mendel 32

Landsteiner, Karl 113
Legally Admissible 24

Marital presumption cases 143
Mitochondrial Inheritance 101
Monohybrid cross 34

mtDNA typing 104

Nucleotides 54

Obligate Paternal Allele 82

Paternity Determination 146
Paternity Index 145
PCR 53
  Amplification 56
  Nucleotides 54
  Primers 55
  Template 54
Probability of Paternity 86, 89
Proteinase K 50
Punnett square 82

RFLP 67
RNA 41
Romanov, Tsar 102

Short Tandem Repeats 71
Siblingship 93
siblingship test 93
Simpson, OJ 51
Single Nucleotide Polymorphisms (SNPs) 42
Southern blotting 68
Standard of Proof 142
STRs 71
Sykes, Bryan 99

Tyler-Smith, Chris 97

Underhill, Peter 97
Uniform Act on Paternity 137

VNTR 67

Watson, James 32

Y-chromosome 94